A Colour Atlas of
Tomato Diseases
Observation, Identification and Control

D. Blancard

INRA Vegetable Pathology Unit
Montfavet, France

Reprinted 1997, 2000

Copyright © 1994 Manson Publishing Ltd
Published 1994 by Manson Publishing Ltd
First published in Great Britain by Wolfe Publishing Ltd, 1992
This edition is published with the help of the French Ministry of Culture

ISBN 1–874545–31–6

First published in French as *Maladies de la tomate* by
INRA,147, rue de l'Université 75007 PARIS and
PHM Revue Horticole, BP 1516 87021 Limoges Cedex
French edition © INRA and PHM Revue Horticole 1988
ISBN 2–7380–0087–8

Printed and bound in Great Britain

For a complete list of Manson titles, please write to
Manson Publishing Ltd, 73 Corringham Road, London NW11 7DL, UK.

Contents

Foreword

This book approaches the subject of diseases of the tomato in a new way, and is original in its structure. It is written by an expert, but in such a way that it can be understood by anyone. The author's originality combines well with his experience, gained over a number of years at a research station. This book, although specialized, is not designed entirely for the specialist. It contains a compilation of information that cannot be found in any other single book. The excellent photographs and illustrations replace complex descriptions of symptoms, so often unintelligible to the non-specialist. The inclusion of summary tables and drawings helps the user to make an accurate diagnosis. The author's experience of examining many thousands of samples from growers and advisers has enabled him to illustrate and describe the essential features of each disease. Considerable emphasis is placed on symptoms of the roots and stems, as it is often failure to recognize these at an early stage that results in disease control failure.

The aim of the book is to determine the cause or causes of disease. The user will easily find the essential information for control treatments where they are feasible or, alternatively, the necessary preventive treatments (e.g., special culture techniques or resistant varieties) that can be used to stop disease occurring in subsequent crops. In addition, the most up-to-date recommendations are given for fungicide treatments.

We must thank the author for having condensed and explained, using clear pictures, the diseases and disorders of the tomato, together with the control measures for these. Other market garden plants will be green with envy at not being treated in this way . . . while the tomato, hearing them, will blush with pleasure.

Paul Rieuf

Acknowledgements

Now that this book is complete, I would like to express my gratitude to all those who contributed to it: first to, Paul Rieuf, who introduced me to the diagnosis of diseases of market garden plants and who has done me the honour of writing the Forward.

I also thank Henri Laterrot, who encouraged my interest in the tomato crop and its diseases. Not only has he written Appendix 2 but also his advice has been of invaluable assistance to me.

In addition, I thank the following for their contribution and comments: Ms Jacquemond, Ms Philouze, and Ms Pichot, Mr Bues, Mr Lot, Mr Molot, Mr Pitrat, Mr Pochard and Mr Rougier.

Dedication

To my parents, my wife Martine, and my two children

3

How to use this book

Before beginning your diagnosis:

1. Select good quality samples (representative of the disease, whole plants with developing symptoms)

2. Collect as much information as possible:
- **On the disease** (distribution in the crop and on the plants – see pages 10 and 11, rate of development, climatic conditions which preceded its appearance or which appear to encourage its spread, etc.)
- **On the plant** (characteristics of the variety, quality of seed, etc.)
- **On the soil** (previous crops, addition of humus or manure, etc.)
- **On the agricultural operations carried out** (method and frequency of irrigation, quantity of water applied each time, application of pesticides to the crops or in the vicinity, etc.)

Diagnosis

1. Locate the symptoms on the diseased plants

	Reference colours in the book	*Where to find the observation guides*
Leaflets and leaves		pages 13, 16, 30, 31, 46, 47, 68, 69
Roots		pages 69, 74, 75, 76, 77
Collar		pages 69, 86, 87, 91
Stem		pages 69, 98, 99, 108
Fruit		pages 120, 121

2. Refer to the part of the diseased plant

At the beginning of each section are:
- The **symptoms studied**
- The **possible causes**

(There are several hypotheses for several symptoms.)

In the case of changes to leaflets and leaves, the wide variety of symptoms have been divided into four sub-sections based on a personal classification of the assessment criteria: **leaflet deformations, leaflet discolorations, spots on leaflets, and withered or dried leaflets.**

3. Select a symptom and turn directly to the pages concerned, or consult all the symptoms of a section

For each symptom there are one or more **possible causes** (several hypotheses correspond to one symptom).

4. Determine the cause of the symptom

To make a choice from the hypotheses presented:
- Compare the symptom or symptoms observed on the plants with those shown on the many photographs.
- use the **additional information for diagnosis.**

Treatment

To combat parasitic micro-organisms, turn to the sheets which form the second part of this book – these are arranged as follows:
- **Symptoms** (photographs showing the symptoms of the disease)
- **Major characteristics of the pathogenic agent** (survival, spread, conditions favourable to development, etc.)
- **Treatment methods** (to be applied during growth and on the next crop).

Information on possible steps to remedy some non-parasitic diseases is also given, in Part Three.

Two **Appendices** conclude the book:

Appendix 1: Reminder of the damage caused by the main pests and some plants parasitic to the tomato.

Appendix 2: List of principal varieties cultivated in the Mediterranean basin with their resistance characteristics. This also gives some general concepts concerning genetic control.

Part One
Diagnosis of Parasitic and Non-parasitic Diseases

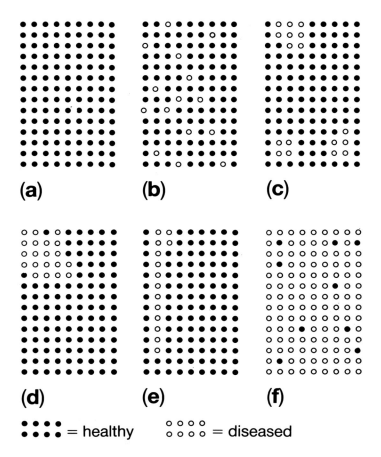

Distribution of plants within the crop

1 (a) Healthy crop.
   (b) Diseased plants spread at random.
   (c) Several small, dispersed groups.
   (d) Very large group.
   (e) Lines of diseased plants.
   (f) Disease affecting most of crop.

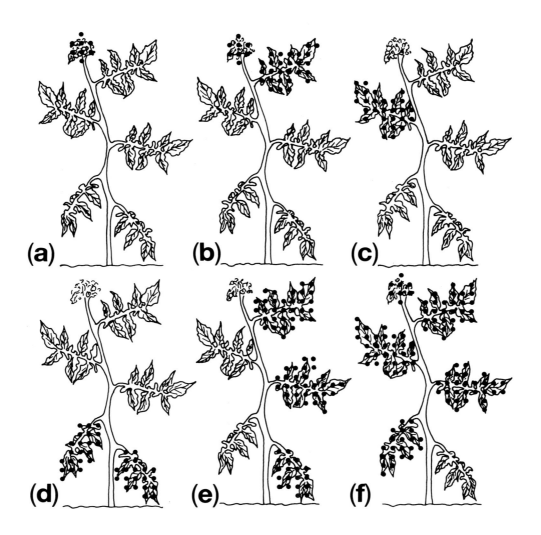

Location of foliar symptom(s) on the plant(s) examined

2 (a) Apex – terminal bud.
   (b) Young leaves (top of plant).
   (c) Patchy and random.
   (d) Old leaves (base of plant).
   (e) Leaves on one side of plant only (unilateral).
   (f) All leaves (general).

1. Irregularities and changes to leaflets and leaves

The symptoms observed on the leaflets and leaves have been divided, for simplicity's sake, into four sub-sections:

- Leaflet and leaf deformations
- Leaflet and leaf discolorations
- Patches on leaflets and leaves
- Wilted and dried leaflets and leaves

OBSERVATION GUIDE

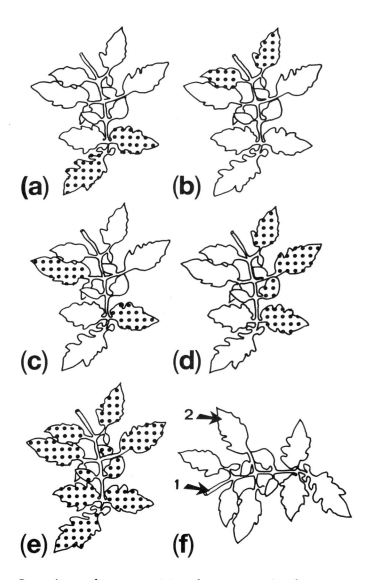

Locations of symptom(s) on leaves examined

3　(**a**) Tip leaflets.
　　(**b**) Base leaflets.
　　(**c**) Some leaflets at random.
　　(**d**) Leaflets on one side of the leaves (unilateral distribution).
　　(**e**) All leaflets (general distribution).
　　(**f**) A leaf (**1** = petiole or rachis, **2** = leaflet).

Leaflet and leaf deformations

SYMPTOMS STUDIED

- Dwarfed, stunted, proliferated growth
- Filiform leaflets
- Partially deformed leaflets
- Leaflets of reduced size
- Wrinkled, rolled, curled leaflets
- Leaflets of reduced size

PROBABLE CAUSES

- Rhabdovirus
- Cucumber mosaic virus
- Tobacco mosaic virus
- Yellow leaf curl virus
- Other viruses
- Stolbur
- Oedema
- Leaf roll
- Sterile mutants, chimeras
- Various phytotoxicities
- Aphids

DIFFICULT DIAGNOSIS

Diseases causing deformations of the leaflets often have several symptoms in common, which makes them difficult to distinguish. We therefore suggest that you look at all the symptoms in this sub-section. These diseases also cause discoloration of the leaflets (it is worth looking at this sub-section too).

Usually, in this situation, there are several hypotheses (with the exception of some diseases with pronounced characteristics).

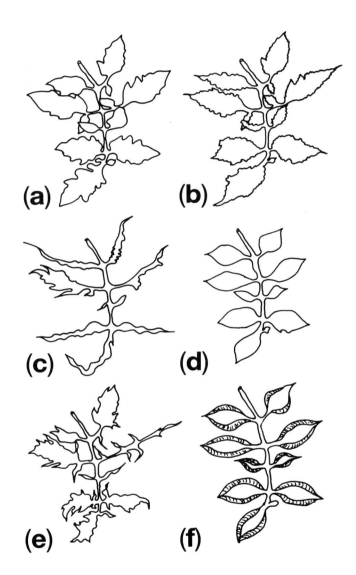

Appearance of some leaflet deformations
4   (a) Normal leaflets and leaf.
     (b) Partially deformed leaflets (serrated at the edge
        of the lamina).
     (c) Highly filiform leaflets.
     (d) Slighly denticulate leaflets of reduced size.
     (e) Highly denticulate leaflets.
     (f) Curled or rolled leaflets.

Examples of leaflet deformations (5 – 12)

5 Filiform leaflets.

6 Serrated leaflets.

7 Reduced size leaflets.

8 Shrivelled leaflets.

9 Proliferation of axillary branches, terminal growth with stiff appearance—stolbur.

11 Numerous leaflets of reduced size, giving the plant a stunted appearance—yellow leaf curl virus.

10 Plant of abnormal appearance, hence its name "crazy plant"—sterile mutant.

12 Plant with early virus attack, reduced in size in comparison with its healthy neighbour —cucumber mosaic virus.

Dwarfed, stunted, and proliferated vegetation

Deformations of leaflets and leaves caused by the diseases described in this section often give an unusual appearance to the entire plant or its apex (temporarily or permanently), which contrasts with that of healthy plants.

In the case of early attack on young plants, their growth may be stopped and they will remain dwarf plants. Their growth could also sometimes be slowed, giving them a stunted appearance.

With later attacks, only the apex and all new growth will be affected, showing the symptoms described in this section.

13   14

13 'Rosette' apex of plant: the very slow, or stopped growth has led to the symptom—rhabdovirus.

14 Numerous curled leaflets of reduced size, giving the plant a bushy appearance—yellow leaf curl virus.

15 15 Mottled, wrinkled leaflets with a tendency to become filiform—tobacco mosaic virus.

17 17 Filiform, wrinkled leaflets—cucumber mosaic virus.

16 16 Filiform leaves, slightly rolled—tobacco mosaic virus.

18 18 Reduced width leaflets, deformed and rolled—sterile mutant.

Filiform leaflets (or leaflets tending to become filiform)

POSSIBLE CAUSES

- **Cucumber mosaic virus (CMV)** (description 26, page 184)
- **Tobacco mosaic virus (TMV)** (description 25, page 183)
- **Sterile mutants (SM), chimeras** (description 31, page 190)

ADDITIONAL INFORMATION FOR DIAGNOSIS

	CMV	TMV	SM
Mosaic at start of attack	+	+	−
Alternate normal and filiform leaves	+/−	+	−
Possibility of symptoms on fruits	+	+	− (Sterile plants)
Type of culture where symptom is observed	Full field Summer Autumn	Sheltered Spring	All
Existence of resistant varieties (list in Appendix 2)	−	+	−
Preventive protection	−	+	−
Distribution in crop, groups	One or more groups at random	In line or general	Very few plants affected (isolated at random)

Sterile mutants
Several types of sterile plant can occur in a crop. The most common type takes the form of plants with reduced, deformed and often rolled leaves, with far fewer long hairs on the stems, leaves and floral peduncles than normal plants. There may also be vigorous plants with very thick stems and leaves. Genetic mutations or chromosome aberrations (in the constitution of chromosomes or their number) cause these symptoms.

19 Highly serrated and blistered leaflets, rather more pointed, where the veins have a tendency to lie parallel—phytotoxicity (excess hormone).

21 Highly denticulate, more pointed leaflets—phytotoxicity (incompatibility of products).

20 Partially deformed and more denticulate leaflets—phytotoxicity (excess fruit setting hormone).

22 Highly denticulate, wrinkled leaflets—chimera.

Partially deformed leaflets

POSSIBLE CAUSES

- **Various viruses**
- **Sterile mutants, chimeras** (description 31, page 190)
- **Various phytotoxicities** (description 31, page 190)
  - —overdose of various pesticides, or pesticides applied in adverse conditions
  - —'hormone' type herbicides
  - —fruit-setting hormones
  - —mixtures of incompatible pesticides
  - —pesticides not suitable for use on the tomato

ADDITIONAL INFORMATION FOR DIAGNOSIS

- **Various viruses** (see pages 21, 23, 33, 35)
- **Sterile mutants** (see pages 18 and 21)
- **Phytotoxicities**

Symptoms developing:

- —fairly rapidly (= relation of cause to immediate effect) after application of a pesticide to the crop or nearby (sprays)
- —later, in the case of poor cultural practices (previous annual or perennial crop cleared with a residual herbicide; perennial crop treated for several years = accumulation of herbicide) or following application of straw or manure from straw made from treated crops

Distribution:

- —applied to plant
  - general
  - start of line
  - near to openings of cloches
  - on one side of plants
- —residual in soil
  - more or less homogeneous, generalized
  - at the ends of greenhouses

Occasional presence of symptoms on fruits—see **268, 291** and **292**.

Did you rinse your sprayer properly? Water from drainage ditches and sometimes wells can be polluted by a herbicide.

See also the other symptoms of phytotoxicities—pages 42, 43, 54, 70, 96, 106.

23

26

24

23 Wrinkled and mottled leaflets—tobacco mosaic virus.

24 Young, slightly denticulate leaflets, blistered and mottled—cucumber mosaic virus.

25 Blistered leaflets, reduced in size and deformed (in shape of a comma) —rhabdovirus.

26 Considerable roll of leaflets at base of plants—leaf roll.

27 More or less curled leaflets – yellow leaf curl virus.

25

27

Wrinkled, rolled, curled leaflets

POSSIBLE CAUSES

- **Rhabdovirus and other viruses** (descriptions 25 TMV; 26 CMV; 29 TYLCV)
- **Leaf roll**
- **Oedema**
- **Aphids**

ADDITIONAL INFORMATION FOR DIAGNOSIS

- **Rhabdovirus and other viruses**

Two **rhabdoviruses** have been observed several times in France:

—one appears to attack only field crops, where several small outbreaks have been observed. In addition to the foliar symptoms (**25**), it also causes a highly characteristic mosaic on the fruits (**240, 299**).

—the other has been found only in sheltered tomato crops in the Perpignan region. It causes the slowing or total cessation of growth of the plants, in which the growing apex is rosetted (**13**).

We have very little information on these viruses at present.

Other viruses can lead to wrinkling and crinkling of leaflets together with discoloration— this is the case in TMV and CMV in particular (**23, 24**).

The **tomato yellow leaf curl virus** sometimes causes a bushy appearance—the symptoms on the leaflets are highly characteristic. Leaflets are often yellow, reduced in size (**11**) and more or less curled (**14, 27**).

- **Leaf roll**

This symptom appears on the leaves at the base of the plant when they are laden with fruit, or when they are subjected to certain climatic or agricultural conditions (prolonged drought, humid and anaerobic soil, severe pruning) which disturbs the water supply to the plants (**26**).

- **Oedema**

Small, highly characteristic blisters are observed on the underside of the leaflets (**86**). In this severe form (shrivelled leaflets) they mainly occur in sheltered crops at the time of sudden changes to night conditions (high humidity, low temperatures), where normally they cause spots (**84, 87**).

- **Aphids**

The presence of small insects (green, black, etc., with little mobility) often grouped into colonies on the underside of the leaflets (see Appendix 1).

28 Reduction in size of new leaflets, the edges of which curl—stolbur.

29 Yellowing and reduction in size of young leaflets—stolbur.

30 Enlarged sepals, green petals of reduced size—stolbur.

31 Incomplete flowers—stolbur.

Leaflets of reduced size

POSSIBLE CAUSES

- **Yellow leaf curl virus (TYLCV)** (description 29, page 187)
- **Stolbur** (description 30, page 189)
- **Other diseases in this sub-section**

ADDITIONAL INFORMATION FOR DIAGNOSIS

- **Yellow leaf curl virus** (see pages 18 and 25)

- **Stolbur**

This mycoplasma disease does not cause mottling. It causes highly characteristic symptoms on the flowers (**30, 31, 54**).

The vegetation of these plants is stunted (**53**). It usually attacks full field crops in summer and autumn. It is very rare under protection – only near openings. See **7, 9,** and **35**.

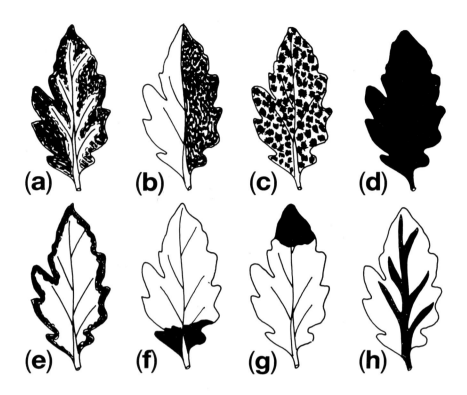

Location and appearance of some discolorations of leaflets

32   (**a**) Between veins.
      (**b**) Unilaterally in relation to the main vein.
      (**c**) Over the entire leaflet, in a diffuse manner.
      (**d**) Over the entire leaflet, general.
      (**e**) At the edge of a lamina.
      (**f**) At the base of the leaflet.
      (**g**) At the top of the leaflet.
      (**h**) Along the veins.

Examples of discoloration of leaflets and leaves (33–40)

33 Mottled leaflets.

34 Silvered leaflets.

35 Blue-tinged leaflets.

36 Leaflets with chlorosis

37 Mosaic on leaves at the apex of a plant—cucumber mosaic virus.

39 Bright yellow patches on the leaflets—tobacco mosaic virus, Aucuba strain.

38 Slightly marbled leaflet—Potato virus Y.

40 Yellow patch mosaic. Alternate light green and dark green patches—tobacco mosaic virus.

Mosaics can have different appearances and intensities

MOTTLED LEAFLETS

POSSIBLE CAUSES

- **Cucumber mosaic virus (CMV)** (description 26)
- **Alfalfa mosaic virus (AMV)** (description 28)
- **Tobacco mosaic virus (TMV)** (description 25)
- **Potato virus Y (PVY)** (description 27)

It is sometimes difficult to assess the presence of a mosaic on the leaflets of a diseased plant—it is practically impossible when they have been exposed to the sun. To observe these, we recommend that you examine the leaflets when illuminated from behind, taking care to shade them.

ADDITIONAL INFORMATION FOR DIAGNOSIS

	CMV	TMV	PVY	AMV
Appearance of mosaic	Visible in green and yellow patches	Highly visible in green, yellow or white patches	Very subtle	With yellowing and necrotic patches starting at the base of the leaflets
Filiform leaflets	+	+	−	−
Symptoms on stems	Elongated brown cankers (sometimes)	−	−	Elongated brown cankers
Symptoms on fruits	+	+	−	+
Existence of resistant varieties	−	+	−	−
Distribution in crop	One or more groups and at random	Generally in lines	One or more groups and at random	One or more groups and at random

41 Cucumber mosaic virus.

43 Cucumber mosaic virus.

42 Tobacco mosaic virus.

44 Tobacco mosaic virus.

These four illustrations of mottling on leaves should give a better idea of what a mosaic is, and show how difficult it is to associate the appearance of a mosaic with one virus rather than another. Although this symptom strongly suggests a virus, it is not adequate for precise identification of the virus responsible. We suggest a specialist laboratory is contacted.

Mottled and necrotic leaflets

POSSIBLE CAUSES

- **Cucumber mosaic virus (CMV)** (description 26, page 184)
- **Potato virus Y (PVY)** (description 27, page 185)
- **Alfalfa mosaic virus (AMV)** (description 28, page 186)

ADDITIONAL INFORMATION FOR DIAGNOSIS

See pages 33 and 51.

Confusion may arise, in particular when seeking the cause(s) of necrotic patches. It is therefore advisable to consult the symptoms of spots and small patches on leaflets (pages 49–53).

45 Inter-vein, patchy yellowing and necrosis on leaflet—potato virus Y, necrotic strain.

46 Inter-vein brown necrotic patches on the upper side of a leaflet—potato virus Y, necrotic strain.

47 Metallic appearance of necrotic patches on the underside of a leaflet—potato virus Y, necrotic strain.

48 Yellowing and necrosis of the base of the leaflets—cucumber mosaic virus or alfalfa mosaic virus.

46

45

47

48

Silvered leaflets with metallic appearance

Silvering (description 31, page 190)

The symptoms that can be seen in **49–52** are highly characteristic of this non-parasitic disease; in addition, certain leaflets are also deformed, and the flowers produced from an affected area are normal in appearance but do not give any fruit. In the case of partial attack, the fruits show silvery green streaks which become pale yellow on maturity. This non-parasitic problem is caused by an anomaly in the development of the tissues under certain conditions: plants from very early sowing or autumn sowing, exposed to daytime temperatures below 18°C.

49 Silvery green 'mosaic' on leaflet—patchy silvering.

50 Metallic appearance of leaflets at apex—tip silvering.

51 Silvery green areas of leaflets and leaflet sectors—silvering.

52 Silvery green stem section—silvering.

Blue-tinged leaflets or leaflets with excessive anthocyanin

POSSIBLE CAUSES

- **Stolbur** (description 30)
- **Phosphorus deficiency** (true or induced)

ADDITIONAL INFORMATION FOR DIAGNOSIS

- **Stolbur** (see page 27)
- **Phosphorus deficiency** (see pages 39–40)

53 Stunted apex of plant, yellowish, excessive anthocyanin—stolbur.

54 Enlarged calix of flowers, giving them a swollen appearance—stolbur.

55 Rolled and blue-tinged leaflets, in particular on the veins—climatic cause (temperature too low).

56 Plantlets with chlorosis and anthocyanin, in particular on the stem—climatic cause (temperature too low).

57 Inter-vein chlorosis of leaflets at the apex of a plant.

58 The veins remain green—iron deficiency.

Deficiencies generally starting with young leaves (apex of plant)

Boron   Slight inter-vein yellowing of leaflets which curl and remain small in size. Subsequently the entire plant is affected.

Calcium   Dull green plants, pale green to yellow at the edge of the lamina. Browning and necrosis of terminal bud, apical necrosis of fruits.

Copper   Stunted plants, rolled leaflets, petioles curved downwards.

Iron   Inter-vein yellowing (to whitening) of leaflets except along veins, which remain green.

Manganese   Inter-vein yellowing of leaflets starting with tissues near to the veins. Deformation and rolling of leaflets.

Zinc   Stunted plants, smaller rolled leaflets with inter-vein yellowing in small patches which might become necrotic.

Yellow leaflets with chlorosis

POSSIBLE CAUSES

- **Nutritional deficiencies**
- **Various phytotoxicities** (description 31)

See also the concept of withered, dried leaflets preceded or accompanied by yellowing.

The yellowing or chlorosis of the leaflets is a symptom frequently observed in tomato cultivation. It may have widely differing distributions:

- Limited to a small area in the form of a spot (see page 57) or linked to a spot in the form of a more or less pronounced yellow halo (see page 49).
- Affecting one side of a leaflet or leaf only. This unilateral yellowing is often characteristic of vascular diseases (see pages 113–114).
- Developing from the veins or between the veins (inter-vein yellowing).
- It may begin with young leaflets or at the apex or the old leaflets at the base of the plants. In some cases, the intermediate leaflets are affected. Sometimes it affects the entire plant. It varies in intensity, sometimes even whitening the leaflets.

This symptom is often characteristic of a nutritional problem in the plants (deficiency, phytotoxicity, etc.) or one or more parasitic attacks either local to the leaves or on other parts of the plant, in particular the roots and stems.

ADDITIONAL INFORMATION FOR DIAGNOSIS

- **Nutritional deficiency**

The term deficiency is often taken to cover both true deficiencies and induced deficiencies.

True deficiencies (elements absent from the soil) are increasingly rare in the soil. Diagnosis of these is difficult as, without exception, the symptoms they produce are discolorations and yellowing of leaflets in varying intensities, which are difficult for a non-specialist to assess. As an example, see pages 38 and 40 which show the principal symptoms of true deficiencies in the tomato.

In most cases, we mean **induced deficiencies** (elements present but not available), which does not make diagnosis any easier. In addition to discovering the nature of the deficiency, and the possibility of confusing the symptoms, the cause or causes must be determined. These can be very diverse, e.g. poor irrigation (too much or too little water), too high or too low a soil temperature or pH, root systems in poor condition, etc.

When faced with these types of symptoms, growers must avoid the temptation of assuming a deficiency without consulting a specialist and carrying out the necessary tests (physical and chemical analyses) on the soil, vegetation, etc.

Deficiencies mainly occur in crops which have been manured haphazardly without any previous soil analysis or nutritional investigation.

59 Leaflets at the base of a yellowing plant.

60 Inter-vein yellowing is homogeneous—
magnesium deficiency.

Deficiencies generally starting with the old leaves (base of plant)

Nitrogen   Weak plants, small pale green leaflets, veins sometimes blue-tinged.

Magnesium   Yellowing of leaflets, starting at edge of lamina and becoming generalized.

Molybdenum   Slight inter-vein yellowing of leaflets which roll up, pale coloration of the finer veins.

Phosphorus   Dull green leaflets, blue tinge on the underside (above all the veins) and on the stem; stunted plants, very thin stems, hollow, poorly coloured fruits. Affects entire plant.

Potassium   Inter-vein patchy yellowing of the leaflets, drying of edges. Softening of fruits.

Sulphur   Slight inter-vein yellowing of leaflets with blue-tinged, necrotic patches. Blue tinge to veins, petioles and stems. Affects the entire plant.

Most induced deficiencies can occur in the following conditions:

- Reduced development or poor state of root system (phosphorus, iron).
- Soil temperatures too low (phosphorus).
- Excessive night temperatures (magnesium).
- Excessive or unbalanced manuring (magnesium if excess potassium, or phosphorus if excess nitrogen or sulphate).
- Low light (phosphorus).
- Poor irrigation (magnesium, calcium or iron if anaerobic, magnesium if lack of water).
- Highly calciferous soil type (phosphorus, iron).
- Plants heavily laden with fruit as harvest approaches.

Various phytotoxicities

61–67 show the nature and distribution of yellowing caused by phytotoxicities. Other symptoms are described on pages 22, 54, 70, 96 and 106.

You will find further information for diagnosis on page 23.

In addition to the illustrations showing different phytotoxicities, the list below describes symptoms observed on tomatoes which have been deliberately contaminated by products from the main groups of herbicides used in agriculture.

- Total blockage of growth or very slow growth, giving the plants an appearance similar to that of stolbur.
- More or less pronounced rolling of the leaflets.
- Filiform leaflets, fern leaf.
- Curved stem.

- Yellowing of base of newly formed leaflets.
- Yellowing and necrosis of tips of leaflets.
- Inter-vein yellowing of leaflets and drying at edges.
- Yellowing of veins, drying of leaflets, cessation of growth.
- Anthocyan production on the underside of leaflets.
- Anthocyan production on leaflet veins.

- Necrosis and drying of edge of lamina.
- Rapid withering and drying of plants.
- Necrotic patches on old leaves.
- Necrotic inter-vein patches, rusty colour.
- Inhibited development of lateral roots.

61 Yellowing of leaflets in patches.

63 Yellowing of leaflets accompanied by necrotic patches.

62 General yellowing of leaflets.

64 Rapid yellowing and whitening of leaflets.

65

66

67

65 Yellowing and whitening starting at base of leaflets.

67 Inter-vein yellowing of leaflets, where the veins remain green.

66 Yellowing starting from veins of leaflets.

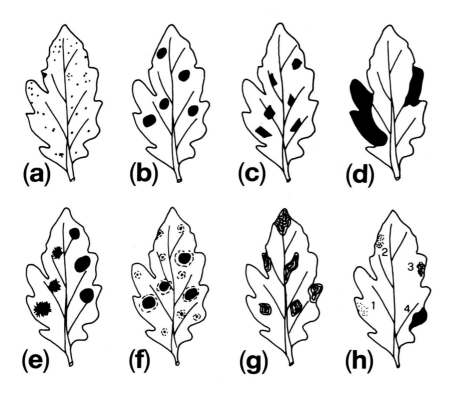

Sizes and shapes of spots on leaflets

68   (**a**) Small spots or specks.
      (**b**) Round spots.
      (**c**) Angular spots.
      (**d**) Extended spots or blotches.
      (**e**) Spots with a diffuse edge (left) or spots with a well-defined edge (right).
      (**f**) Spots surrounded by a halo.
      (**g**) Spots in ring form or as concentric rings.
      (**h**) Combination of different forms of spots.

69

70

71

72

Examples of spots on leaves (69–76)

69 Small spots or specks.

70 Concentric ring spots.

71 Yellow spots.

72 Oily spots.

73 Brownish black, angular spots on young leaflets—*Xanthomonas campestris* pv. *vesicatoria*.

75 Brown, clearly outlined angular spots with a clearly defined yellow halo, fused together in places—*Pseudomonas syringae* pv. *tomato*.

74 Brown spots with a discrete yellow halo (fused together in places), the centre of which may fall out—*Xanthomonas campestris* pv. *vesicatoria*.

76 Brown spots along the stem—*Xanthomonas campestris* pv. *vesicatoria*.

Brown spots on leaflets

POSSIBLE CAUSES

- *Pseudomonas syringae* pv. *tomato* (description 1, page 152)
- *Xanthomonas campestris* pv. *vesicatoria* (description 2, page 152)
- **Alfalfa mosaic virus (AMV)** (description 28, page 186)
- **Potato virus Y (PVY)** (description 27, page 195)
- **Virus complex**
- **Various phytotoxicities** (description 31, page 190)

ADDITIONAL INFORMATION FOR DIAGNOSIS

	Pseudomonas tomato	*Xanthomonas vesicatoria*	Potato virus Y	Phytotoxicities
Spots spread over leaves and plant	+ or − Localized	+ or − Localized	+ or − General	See page 23
Mottle may accompany spots	−	−	+	−
Symptoms on stems	+	+	−	+/−
Symptoms on fruits	+	+	−	−
Rapid development if high humidity (rain, sprinkling)	+	+	−	−
Attack in spring	+	−/+	+	+
Attack in summer, autumn	+	+	+	+

77

78

79

- *Pseudomonas syringae* **pv. tomato**
- *Xanthomonas campestris* **pv. vesicatoria**

These two bacteria cause almost identical symptoms on all structures of the tomato except the fruit. It is therefore advisable, when brown spots are found on the leaflets, to observe the fruits carefully, where the symptoms are highly characteristic.

77 Brown spots on sepals—*Xanthomonas campestris* pv. *vesicatoria*.

78 Brown spots on calyx and floral peduncle—*Pseudomonas syringae* pv. *tomato*.

79 Spots on fruits.

Distinguishing symptoms

Corky pustules (4–5 mm in diameter) surrounded by an oily halo (left)—*Xanthomonas campestris* pv. *vesicatoria*.

Small brown spots (2–3 mm in diameter) or specks, very superficial (right)—*Pseudomonas syringae* pv. *tomato*.

80 Diffuse brown necrotic spots (start of attack)—potato virus Y (necrogenic strain).

81 Diffuse brown spots covering entire leaflet accompanied by chlorosis—potato virus Y (necrogenic strain).

82 Brown necrosis localised on veins and adjacent tissues—virus complex (TMV and PVY).

- **Alfalfa mosaic virus**
- **Potato virus Y**
- **Virus complex**

See symptoms of mottled/necrotic leaflets, page 35.

83

84

85

86

87

Other small spots on leaflets

POSSIBLE CAUSES

- *Clavibacter michiganensis* subsp. *michiganensis* (description 3, page 153)
- *Penicillium* sp.
- *Stemphylium spp.* (description 11, page 163)
- **Oedema** (description 31, page 190)
- **Various phytotoxicities** (description 31, page 190).

ADDITIONAL INFORMATION FOR DIAGNOSIS

- **Stemphylium spp.**

Several species of *Stemphylium* cause the same symptoms (small brown spots which lighten, crack and dry out) on the leaflets of the tomato: *Stemphylium solani*, *Stemphylium floridanum* (or *S. lycopersici*), *Stemphylium botryosum* f. sp. *lycopersici* and *Stemphylium vesicarium*. The development of *Stemphylium* is encouraged by warm, moist weather (in particular, water on the plants). This disease (stemphyliosis) often causes serious damage in association with either *Pseudomonas syringae* pv. *tomato* or with *Alternaria dauci* f. sp. *solani*, making identification difficult.

Many varieties have a gene with effective resistance against these four species (see list of varieties in Appendix 2).

- **Oedema**

The small blisters are highly characteristic of this non-parasitic disease, which occurs with sudden changes in the day/night climate (high humidity, low temperatures). The blisters occur in high numbers, with shrivelling and rolling of the leaflets.

Protected crops are most affected (late winter, early spring, late autumn).

Plants with symptoms can be dispersed over the entire crop or they may be localized near the coldest and most humid areas of the greenhouse. The disease is always linked with excessive humidity in the greenhouse.

---

83 Small, slightly angular spots (from 2–10 mm in diameter) with chlorosis at the edge. Spots are initially brown but with the centre becoming lighter (grey) and cracking—*Stemphylium vesicarium*.

84 Small spots (2–4 mm in diameter), initially yellow then beige, on the upper side of the leaflet—oedema.

85 Slightly angular brown spots, cracked in the centre, chlorosis of part of the lamina—*Stemphylium vesicarium*.

86 White blisters (equals oedemas) on the underside of a leaflet, sometimes localized along the veins—oedema.

87 'Oedemas' that have burst and scarred, beige in colour—oedema.

88

89

90

● **Various phytotoxicities**

Overdoses of certain pesticides sometimes cause small necrotic spots on the leaves (88) which may be confused, for example, with those caused by the necrotic strains of potato virus Y. The spots are caused by local 'scorch' of foliar tissues; this occurs particularly when plants are accidentally treated with fertilizer particles (89) or splashes of contact herbicide (90). Often, only the exposed leaves or those on one side of the plant are affected. On these, the symptoms are particularly pronounced where the product(s) may have been retained longer.

See also pages 23, 41, 70, 96, 106.

91 Small round beige spots on the upper side of the leaflets—*Penicillium* sp.

92 On the underside of the leaflets, white fructifications on spots indicate a *Penicillium* developing on the larva of a whitefly—*Penicillium* sp.

● ***Penicillium* sp. (a curiosity)**
Where biological control is used (*Encarsia formosa—Trialeurodes vaporariorum*), small beige spots are often found on the leaflets (Figure 91). These are caused by a *Penicillium*, which initially uses the larvae of the whitefly as a nutritional base to grow on the leaflets before colonizing the lamina.

Growers should see this phenomenon as a 'necessary evil' which does not cause any damage, linked to biological control.

93 Small cankerous white spots—*Clavibacter michiganensis* subsp. *michiganensis*.

● ***Clavibacter michiganensis* subsp. *michiganensis***
This bacterium may be the cause of cankerous spots (**93**) initially white then beige to dark brown on leaflets. This symptom is very rare—it indicates a serious outbreak and is always accompanied by withering of the leaflets, browning of the stem vessels, etc. (see pages 109–111).

● **Upper side of leaflets** (for underside see page 58)

94 Bright yellow spots, sometimes angular—
Leveillula taurica.

96 Pale yellow diffuse spots—*Fulvia fulva.*

95 Patch turning necrotic in centre, with a
white down—*Leveillula taurica.*

97 Yellow leaflet, completely colonized,
covered with olive–brown fructifications of a
fungus—*Fulvia fulva.*

Yellow spots on leaflets

POSSIBLE CAUSES

- *Erysiphe* sp (description 9, page 161)
- *Fulvia fulva* (*Cladosporium fulvum*) (description 8, page 160)
- *Leveillula taurica* (description 9, page 161)
- **Hyperparasitic fungi of *Fulvia fulva* (*Acremonium sclerotigenum, Hansfordia pulvinata*)**

ADDITIONAL INFORMATION FOR DIAGNOSIS

	Fulvia fulva	*Leveillula taurica*
Felting on underside of leaves (page 58)	Clearly visible. Initially white, then turning olive–brown	Often invisible. Always white when present
Existence of resistant varieties (see list in Appendix 2)	+	–

98

99

100

101

● **Underside of leaflets** (for upper side see page 56)

98 Yellow spots, sometimes with light brown dots (making it look 'dirty')—*Leveillula taurica*.

99 Pale yellow spots covered with a slight white down—*Leveillula taurica*.

100 Leaflets attacked by two fungi at a time. The more angular spots of *Oidium* remain white. The rounder spots caused by *Fulvia fulva* turn dark brown as the spore develops.

101 More diffuse yellow spots covered with spores, white at first then olive–brown, of *Fulvia fulva*.

102

103

• A new powdery mildew on tomato (Erysiphe sp.)

Symptoms of powdery mildew on tomatoes (**102**) (similar to those on cucurbitaceae) have been observed in greenhouses in Northern European countries (England, Netherlands). This disease is not caused by *Leveillula taurica*, but an *Erysiphe* sp. as yet unidentified. *Erysiphe* spp. have already been reported on tomatoes from elsewhere: *Erysiphe cichoracearum* and *Erysiphe polygoni*.

This disease was not present in France until very recently; in fact it was introduced into the Orléans region through young tomato plants imported from Holland. It appears to develop in protected crops at times (spring) when *Leveillula taurica* is not found.

104 Colonies of *Acremonium sclerotigenum* developing on the sporing patches of *Fulvia fulva*.

104

102 Powdery white spots on the upper side of leaflets—*Erysiphe* sp.

103 Cladosporiosis spots fully covered by the mycelium and white fructifications of *Hansfordia pulvinata*.

• Hyperparasitic fungi of *Fulvia fulva* (a curiosity)

In certain greenhouses, the development of white colonies is sometimes observed on the olive–brown felty spots caused by *Fulvia fulva*. Two antagonistic fungi can attack *Fulvia fulva*: *Hansfordia pulvinata*, already well known and perhaps soon to be available commercially to counter cladosporiosis; and *Acremonium sclerotigenum*, recently discovered.

These two fungi are not sufficiently effective in their natural state to prevent the development of cladosporiosis, but they do help reduce the inoculum present under protection.

105 Beige spots in concentric rings, wrinkled on the top of a leaflet—*Botrytis cinerea*.

107 Small irregular brown spots with chlorosis at their edges, on leaflets—*Alternaria dauci* f.sp. *solani*.

106 Large light brown spots with darker concentric rings on the leaflet—*Botrytis cinerea*.

108 Small brown spots consisting of concentric rings, resembling a target—*Alternaria dauci* f.sp. *solani*.

Concentric circular or curved spots on leaflets

POSSIBLE CAUSES

- *Alternaria dauci* **f.sp.** *solani* (description 6, page 157)
- *Botrytis cinerea* (description 7, page 158)
- **Indeterminate, non-parasitic problem**

ADDITIONAL INFORMATION FOR DIAGNOSIS

	Alternaria dauci f.sp. *solani*	*Botrytis cinerea*
Frequency of attack	+/−	+++
Under protection	+/−	+++
Field grown	+	−/+
Early crops	−	+++
Full season crops	+	+
Development of epidemic	Slow	Fast

The symptoms are sufficiently characteristic for reliable diagnosis. See also the illustrations on pages 62 and 63.

109 Translucent to white rings (3–6 mm in diameter) on fruit, with surrounding tiny spots resembling 'haloes'—*Botrytis cinerea* (ghost spots).

111 Old leaflet covered with rounded brown spots, from which yellowing begins—*Alternaria dauci* f.sp. *solani*.

110 Beige wrinkled spot, flame-shaped, developing from the edge of the leaflet—*Botrytis cinerea*.

112 Necrotic sepals, concave spots (recessed) at the peduncle attachment of the fruit, with a black mould—*Alternaria dauci* f.sp. *solani*.

113 Brown peduncles and flowers covered with a characteristic grey mould—*Botrytis cinerea.*

115 Small brown elongated patches on the stem, lighter at the centre (start of attack)—*Alternaria dauci* f.sp. *solani.*

114 Dark brown discoloration on the rachis, covered with grey felting—*Botrytis cinerea.*

116 Brown spots on the rachis, with grey centre—*Alternaria dauci* f.sp. *solani.*

117

118

117 Leaflet covered with numerous zoned brown spots, with metallic coloration (under protection).

118 Dark brown to brown spots and areas with concentric 'rings', yellowing and anthocyanization of certain leaflets (field crops).

● **Indeterminate non-parasitic problem**

On several occasions, foliar spots (**117, 118**) have been observed on tomato plants in numerous crops, both under protection and in the field. Often, these spots have been mistaken for those of *Alternaria dauci* f.sp. *solani* (in contrast to alternariasis, there are no symptoms on the stem and fruit). No pathogen (bacterium, fungus, virus) has been isolated on these plants. These spots have always appeared at the same time in several crops located in the same region, sometimes several miles apart. In the affected crops, most of the plants have these spots: on one plant, almost all the leaflets were affected. In the days following the outbreak, the plants grew normally and the new leaflets did not show the symptom.

It would appear that this is a non-parasitic disease, probably physiological. A cause has not yet been identified, but it is hoped that this information will help avoid useless treatment if this problem is encountered, and growers can be reassured as to the future of their crops.

Spots of oily appearance on leaflets

POSSIBLE CAUSE

● *Phytophthora infestans* (late blight) (description 10, page 162)

ADDITIONAL INFORMATION FOR DIAGNOSIS

Blight also causes patches or brown spots on stems, even encircling them. It can sporulate on leaves in the form of a white down.

Its attacks are often sudden and generally occur during cloudy and humid weather. The systematic application of fungicides to crops means that this disease is much less common than in the past.

Initial attacks of *Phytophthora nicotianae* var. *parasitica* (soil blight) on fruits have been mistaken for blight (see illustrations on pages 135 and 138 for a clear distinction). Blight does not cause any symptoms on the fruit without the leaflets and leaves being affected.

If symptoms are observed only on the leaves, namely several dried leaflets, **also see the sub-section on withered and dried leaflets and leaves** (page 67 onwards).

120

121

119 Young 'blighted' fruit, brown and partly covered with a white felting—*Phytophthora infestans*.

122

120 Considerable part of leaflet (area) pale and oily, browning of some veins—*Phytophthora infestans*.

121 Large patch on leaflet drying out in centre, oily and pale at edge—*Phytophthora infestans*.

122 Marbled brown spots, irregularly bumpy on surface—*Phytophthora infestans*.

119

Examples of wilting and drying of leaflets

123 Wilted leaflets.

125 Yellowing and drying leaflets.

124 Partially wilted leaflets.

126 Dried leaflets.

127

**With LEAFLET WILTING
Be careful!**

Also check

28

130

129

**The STEM
(exterior and
interior)**

The ROOTS

The COLLAR

If your observations are conclusive, refer to the organ or part of plant concerned.

131 Inter-vein wilting in spots—phytotoxicity (herbicide).

132 Rolling of partially dried leaflets— phytotoxicity (herbicide).

133 Necrosis and drying of edge of leaflets, which are rolling in—phytotoxicity (excess salinity).

134 Inter-vein drying of leaflets— phytotoxicity (herbicide).

● **Various phytotoxicities**

If no changes have been observed in the parts of plants already described, it should be remembered that several **herbicides** can be responsible for foliar necrosis and sudden wilting or desiccation, **131–134** (see pages 23 and 41 for more information).

● **Russet mite**

The mite *Aculus lycopersici* (responsible for russet mite symptoms) causes fairly characteristic drying of leaflets at their base. This pest is often incorrectly diagnosed, at least in the early stages of attack (**203** and Appendix 1).

● **Various viruses**

Several viruses, in particular the necrotic strains of cucumber mosaic virus and alfalfa mosaic virus, can be the cause of necrotic lesions on the leaflets which develop reasonably quickly and cause drying of all or part of the plant (**48, 135**). Often this drying of leaves is accompanied by long superficial brown areas on the stems and petioles (**201, 202**) and various changes to the fruits (**276, 277**).

● **Phytophthora infestans**

These viruses should not be mistaken for late blight, for which the fungus responsible (*Phytophthora infestans*) causes very similar symptoms (in particular, the oily spots on the leaflets, which dry quickly); see **120, 121**.

135

135 Necrosis and drying of a leaf, and a brown area on the stem beginning at the axil of a petiole—cucumber mosaic virus (necrotic strain).

2. Irregularities and changes to the roots

SYMPTOMS STUDIED

- **Yellowing, browning of rootlets and roots**
- **Root rot**
- **Corky roots and/or roots covered with small black spots**
- **Galls and growths on roots**

POSSIBLE CAUSES

- *Agrobacterium* **sp.**
- *Colletotrichum coccodes*
- *Fusarium oxysporum* **f.sp.** *radicis-lycopersici*
- *Phytopthora* **spp. and** *Pythium* **spp. = various pythiaceae**
- *Pyrenochaeta lycopersici*
- *Rhizoctonia solani*
- *Spongospora subterranea*
- *Meloidogyne* **spp.**
- **Root asphyxia**
- **Excess salinity**

It is recommended that you also refer (if briefly) to the section on 'irregularities and changes of the collar', because amongst the micro-organisms listed above, some also attack the collar, possibly at the same time.

RELATIVELY EASY DIAGNOSIS
(but create the correct conditions for diagnosis)

The root system is the least known part of the plant: many growers and scientists do not know how to assess its condition, as often they have not created the right conditions for doing so.

- First, **collect them carefully,** avoiding sudden jerks, or else the altered, weakest parts (but the most interesting for diagnosis) will remain in the soil.
- Then **wash thoroughly in water** to remove particles of soil which frequently mask certain symptoms.

Now you can see the roots, examine them very carefully.

Wash roots well

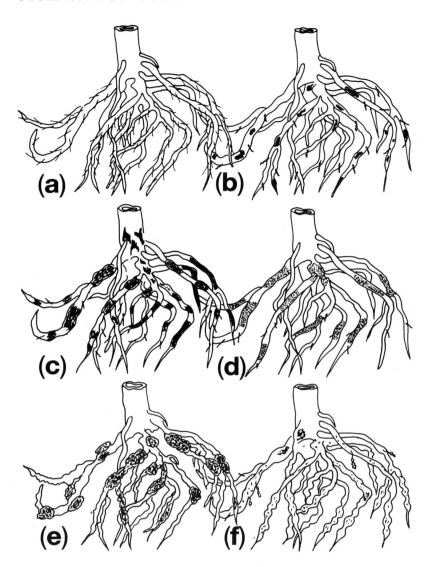

Irregularities, changes to roots

136   (a) Healthy roots with rootlets
     (b) Necrotic roots, brown rootlets
     (c) Corky sleeves on roots
     (d) Black spots on roots
     (e) Growths on roots
     (f) Galls on roots

Examples of irregularities and changes to roots (in comparison with healthy roots)

137 Healthy roots.

139 Yellow and brown roots.

138 Roots with galls.

140 Corky roots.

Location of principal underground parts of the tomato

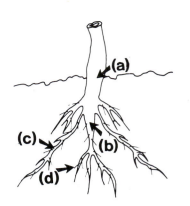

141   (**a**) Collar
      (**b**) Main root (= taproot)
      (**c**) Secondary root
      (**d**) Rootlet

142

143

144

145

142 Browning of all roots contained within a cube of rockwool.

143 Small necrotic lesions, dark brown to brown, on the roots.

144 Browning and destruction of the cortex (E) of the root; only the central cylinder (C) (stele) survives in parts.

145 Browning of central cylinder (C) (stele) of the root, with the cortex (E) remaining healthy.

76

Yellowing and browning of rootlets and roots

POSSIBLE CAUSES

All diseases responsible for irregularities and changes to the roots (descriptions 13–18)

All diseases attacking roots cause a diffuse yellowing and/or browning (localized or generalized), necrosis and disappearance of numerous rootlets. In the most serious cases, the root system may be totally destroyed. The vessels of the tap root and collar turn yellow and then brown (**145**).

Some diseases (corky root disease, etc.) also cause highly characteristic changes to the roots, making their identification very easy, as will become apparent on the following pages.

Others are more difficult to identify just from the symptoms listed above, and **laboratory tests are essential** to determine the cause(s). Most root diseases are caused by fungi: *Rhizoctonia solani* and fungi in the **Pythiaceae** (several species of *Phytophthora* and *Pythium*), or cultural problems leading to **root asphyxiation** or **scorch** following excessive salt concentrations (**excess salinity** after over-use of fertilizer or drying of the substrate).

These problems often occur at the same time, or more precisely, some stimulate the occurrence of others. For example, attacks by fungi in the Pythiaceae are more serious if the plants have been over-watered. In contrast, *Rhizoctonia solani* can cause problems in dry soils.

It should be noted *Rhizoctonia solani* and *Phytophthora nicotianae* var. *parasitica* also cause collar cankers (see pages 88 and 89).

(Consult a specialist laboratory if possible.)

Sometimes, by giving honest replies to the following questions, the problem can be diagnosed (an affirmative response to one of these questions will confirm it):

- Have the tomatoes been over-watered or very cold water used to irrigate them?
- Are the diseased plants located in the wetter parts of the soil?
- Were they planted out at a time when the ground was still cold and damp?
- Was excessive irrigation used during propagation or when planting out?
- Is the soil well drained?
- Has too much fertilizer been added before planting or during culture?

Symptoms characteristic of a root problem, but insufficient to identify the cause

146 Browning of the vascular tissue located at the collar in comparison with healthy vessels.

146

47

149

148

147 Browning of roots, their central cylinder (steles) and the vascular tissue of the tap root and collar—*Fusarium oxysporum* f.sp. *radicis-lycopersici*.

149 Brown canker, slightly sunken, developing on one side of the collar and stem only (flame-shaped)—*Fusarium oxysporum* f.sp. *radicis-lycopersici*.

148 Reduced root system, brown and rotted, with chocolate brown vascular tissue in the lower area of the stem—*Fusarium oxysporum* f.sp. *radicis-lycopersici*.

These changes of course lead to wilting of the leaflets and leaves at the apex and/or a yellowing of leaves at the base of the plants. These often occur as harvest approaches, when the plants are heavily laden.

These symptoms can be confused with those caused by asphyxia or attack by the various Pythiaceae fungi, and sometimes by *Fusarium oxysporum* f.sp. *lycopersici* (see table on page 79).

Fairly soon after root attack, the vascular tissue turns brown, in particular in the tap root and collar. This browning then spreads to the vascular tissue in the stem. Not all of this tissue is affected; often, a 'brown thread'—a brown line a few millimetres wide —can be found extending upwards in the stem for a few centimetres. The browning may also affect a large proportion of the stem's vascular tissue, and can be found up to at least 30 cm above the collar.

Root rot

POSSIBLE CAUSES

- *Fusarium oxysporum* f.sp. *radicis-lycopersici* (description 16, 170)
- **Other highly developed root diseases**

ADDITIONAL INFORMATION FOR DIAGNOSIS

This parasitic fungus starts on roots where it causes numerous dark brown necroses leading to a generalized browning of the roots and their total destruction (rot).

When the disease is well developed, a brown moist canker appears at collar level, slightly depressed and clearly outlined, sometimes covered with a pale pink to salmon mould (fungi fructification, see **168**). These cankers have a characteristic shape because they develop on one side of the collar or stem, resulting in a flame-like shape. Under very humid conditions the cortical tissues of the collar rot completely and become detached, and the affected plant appears to be strangled.

	Fusarium oxysporum f.sp. *radicis-lycopersici*	Various pythiacious fungi Root asphyxia	*Fusarium oxysporum* f.sp. *lycopersici* (Fusarium wilt)
Alteration of root system	+ +	+/+ +	−/+
Browning of vessels:	+ + +	+/−	+ +
			(**210, 220**)
Appearance	Chocolate brown	Yellow to light brown	Dark brown with darker brown lines
Location on plants	Tap root, collar, to 30 cm above collar	Tap root, collar	Very high in the stem frequently in excess of 30 cm
Unilateral yellowing of leaves and leaflets at base of plants	−	−	+
Yellowing of part of stem	−	−	+
Possible presence of collar canker	+	+/−	−
Salmon pink mould on canker	+	−	−

150

151

152

150 Numerous small necrotic lesions, very brown parts of roots, corky in places—*Pyrenochaeta lycopersici*.

151 Dark brown corky zones, cracked in places—*Pryenochaeta lycopersici*.

152 Light or dark brown roots with tiny black spots—*Colletotricum coccodes* (to be observed with a magnifying glass).

153 Very corky and swollen root surface—*Pyrenochaeta lycopersici*.

153

Corky roots and/or roots covered with small black spots

POSSIBLE CAUSES

- *Colletotrichum coccodes* (description 14, page 168)
- *Pyrenochaeta lycopersici* (description 13, page 166)

ADDITIONAL INFORMATION FOR DIAGNOSIS

Observation of highly typical symptoms is sufficient for diagnosis. Note that the two fungi often attack at the same time.

See also **140**, which shows characteristic corky sleeves caused by *Pyrenochaeta lycopersici*.

154

155

157

156

154 Small round galls on roots—*Meloidogyne* sp.

155 Roots with numerous swellings—*Meloidogyne* sp.

156 Numerous small white protuberances on roots—*Spongospora subterranea* (soil-less).

157 Bumpy galls, sometimes corky and moist, on roots—*Spongospora subterranea* (soil-less).

Galls and growths on roots

POSSIBLE CAUSES

- *Agrobacterium* sp. (description 5, page 155)
- *Spongospora subterranea* (description 15, page 169)
- *Meloidogyne* spp. (description 18, page 173)

ADDITIONAL INFORMATION FOR DIAGNOSIS

The highly characteristic symptoms of these three plant diseases make identification easy.

Changes caused by gall nematodes (*Meloidogyne spp.*) are frequently observed on tomato roots, in particular in soils in which tomatoes or leguminous crops have been grown for several seasons. It is not unusual to find them in association with *Pyrenochaeta lycopersici*.

Spongospora subterranea is responsible for the powdery galls on potato tubers, but is rarely found on tomato roots. The numerous galls it can cause, although very spectacular (in particular in soil-less culture in peat, perlite or rockwool, etc.) have little effect on the development of the plants. The galls often appear to be located in the root ball. This unusual distribution may result from contamination during propagation or excessive wetness in the ball caused by the position of the drip irrigation.

Agrobacterium sp. (bacteria) has also been isolated from root galls such as those shown in **156** and **157**. Its pathogenicity has not been proven.

159

158 Unusual gall, rough and corky, brown in colour—*Spongospora subterranea* (in soil).

159 Swollen root, dark brown, warty surface —*Spongospora subterranea* (in soil).

158

3. Irregularities and discoloration of the collar (and underground stem)

SYMPTOMS STUDIED

- **Discoloration and brown necrosis of the collar**
- **Light brown to beige discoloration of the collar**
- **Corky collar and tap root**

POSSIBLE CAUSES

- *Botrytis cinerea*
- *Didymella lycopersici*
- *Fusarium oxysporum* f.sp. *radicis-lycopersici*
- *Phytophthora nicotianae* var. *parasitica*
- *Rhizoctonia solani*
- *Sclerotinia sclerotiorum*

- **Asphyxia of the collar**
- **Excess salinity**

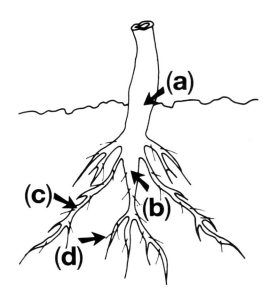

Location of major buried parts of the plant

160   (**a**) Collar (junction of stem and main root).
      (**b**) Main root (= tap root).
      (**c**) Secondary root.
      (**d**) Fine lateral roots

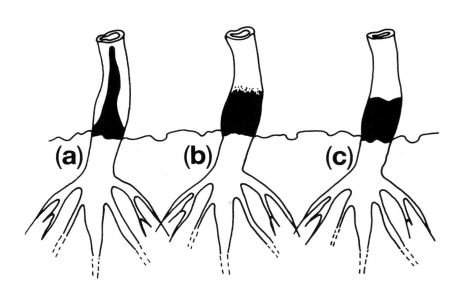

Appearance of some cankers

161   (**a**) Flame-shaped canker.
      (**b**) Diffuse edge canker.
      (**c**) Clearly outlined canker.

Examples of discoloration of the collar (162–167)

162 Brown and moist rot of collar and buried part of stem.

163 Brown rot, wet at collar, disappearance of part of the cortex.

162

163

164 Dark brown to black lesion, moist, diffuse at leading edge—*Phytophthora nicotianae* var. *parasitica*.

166 Brown discoloration, clearly outlined, dry in appearance—*Rhizoctonia solani*.

165 Brown wet rot of collar of recently planted out plants—*Phytophthora nicotianae* var. *parasitica*.

167 Black canker, clearly outlined on the collar—*Didymella lycopersici*.

Discoloration and brown necrosis of the collar

POSSIBLE CAUSES

- *Didymella lycopersici* (description 19, page 174)
- *Fusarium oxysporum* **f.sp.** *radicis-lycopersici* (description 16, page 170)
- *Phytophthora nicotianae* **var.** *parasitica* (description 20, page 175)
- *Rhizoctonia solani* (description 21, page 177)

ADDITIONAL INFORMATION FOR DIAGNOSIS

	Didymella lycopersici	*Fusarium oxysporum* f. sp. *radicis-lycopersici*	*Phytophthora nicotianae* var. *parasitica*	*Rhizoctonia solani*
Canker on collar	+/− Clearly outlined, brown, moist at start of attack, light brown sometimes at end of attack. Black spots	+/− Clearly outlined, recessed, pink 'mould'	+ Diffuse at edges, moist	+/− Clearly outlined, rather dry
Discoloration of roots	−/+	++	+/−	+/−
Marked browning of vessels, at some distance from canker	−/+	++	+/−	−/+
Canker starting from pruning wounds	+ Under protection mainly	−	−	−
Appearance stage of symptoms	Rare in propagation, field crops	Mainly near to harvest	Propagation, after planting out. Later in soil-less cultures	Propagation, after planting out
Additional symptoms, see	188, 192	144, 148, 149, 234	142	

168

169

170

168 Brown canker, moist, clearly outlined, recessed (when pressed), covered with salmon-pink mould—*Fusarium oxysporum* f.sp. *radicis-lycopersici*.

170 Browning of vessels, cortical tissues, tap root and collar (particularly pronounced at base of rotted roots)—*Fusarium oxysporum* f.sp. *radicis-lycopersici*.

169 Cortical brown-pink lesion reaching several centimetres above collar—*Fusarium oxysporum* f.sp. *radicis-lycopersici*.

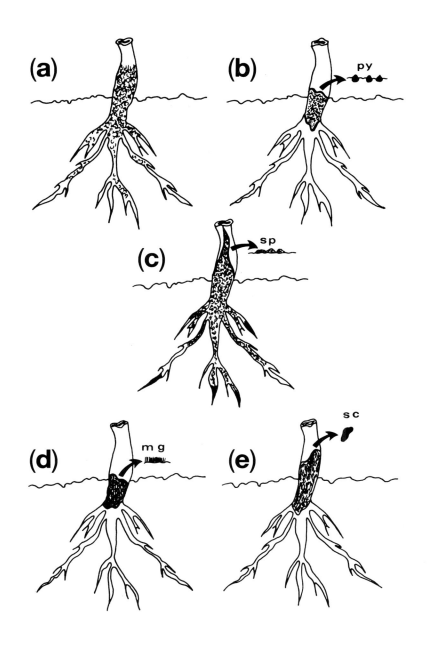

171 Discoloration of the collar caused by parasitic fungi

	Fungus	Characteristic structures
(a)	*Phytophthora nicotianae* var. *parasitica*	Sporangia or spores in tissues
(b)	*Didymella lycopersici*	Pycnidia (py) with the appearance of black spots
(c)	*Fusarium oxysporum* f.sp. *radicis-lycopersici*	Sporodochiae (sp) resembling a salmon-pink mould
(d)	*Botrytis cinerea*	Grey conidial sporulation called grey mould (mg).
(e)	*Sclerotinia sclerotiorum*	Black sclerotia (sc) of varying size, present mainly in the pith

172 Light brown to beige discoloration, brown at edges on young plant (having developed from cotyledons)—*Botrytis cinerea*.

173 Beige to light brown canker, dry, at collar of young plant—*Botrytis cinerea*

174 Beige, dry canker at the base of the stem; black sclerotia in the pith—*Sclerotinia sclerotiorum*.

175 Brown discoloration of collar, covered with white felting and a sclerotium—*Sclerotinia sclerotiorum*.

Light brown to beige discoloration of the collar

POSSIBLE CAUSES

- **Botrytis cinerea** (description 7, page 158)
- **Sclerotinia sclerotiorum** (description 22, page 178)

ADDITIONAL INFORMATION FOR DIAGNOSIS

These two fungi are normally easy to identify, and their fructifications on plant tissues are highly characteristic (grey mould = *Botrytis cinerea;* sclerotia + white felting = *Sclerotinia sclerotiorum*). If these are not present, their development can be encouraged by placing the damaged parts of the collar or stem in a plastic bag or a sealed container with a moist tissue. Under these humid conditions, the two fungi will 'fructify' fairly rapidly.

- *Sclerotinia sclerotiorum* rarely causes damage to the collar of the tomato; it develops more frequently on adult plants, particularly from pruning wounds (see **189, 191**).

- *Botrytis cinerea* is much more common, and can attack both during propagation and after planting out. On young plants (planted too deeply perhaps) it readily colonizes their senescent cotyledons. It then spreads to the stem and encircles it quickly. It can also occur after planting, causing cankers on the collar or on parts of the stem buried too deeply at planting.

176 Corky, scarred crack on the collar.

178 Collar with large diameter caused by thickening of the cortical tissues; breakage and disappearance of tap root.

177 Yellow to light brown corky tap root with a longitudinal crack.

179 Brown corky collar, constricted tap root, browning and corkiness of roots nearest to the tap root.

Corky collar and tap root

Corkiness of the stem base is frequently observed in crops; its cause is not known. These symptoms and their development resemble those described in pimentos by the Italians, under the name 'basal necrosis of the pimento':

- Slight yellowing and wilting, often reversible, of leaflets and leaves.
- Necrosis and corkiness of the collar (the diameter of which may be large) and tap root (main root), which sometimes also show longitudinal cracks (caused by lack of elasticity of scar tissue during the growth of the tap root and collar). Where this is the case, a browning of the vessels and/or pith may occur locally.
- Reduced diameter of tap root in relation to that of the collar and stem.
- Necrosis and corkiness of large secondary roots.
- When this basal symptom is well developed, numerous roots are destroyed. Sometimes the tap root will 'break' very easily.

As with the pimento, the cause of these symptoms does not appear to be parasitic but linked to 'micro-asphyxia', which occurs in particular very early (during propagation) and is repeated with time. Its development is more likely when cultural practices result in water stagnation for brief periods, and when very cold water is used. This repeated 'micro-asphyxia' of plants causes necrosis of submerged parts and the formation of corky scar tissue. Excess salinity can also be an encouraging factor in its appearance. Often, many plants show the same symptoms.

NB: *Pyrenochaeta lycopersici*, during very severe attacks, can also cause corkiness of the collar (but in this case, many corky sleeves are also present on the roots) (see pages 80–81).

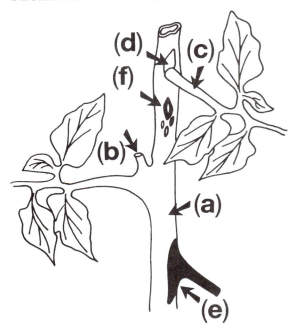

What can you see on the outside of the stem?

181 (**a**) Stem.
  (**b**) Pruning wound.
  (**c**) Petiole or rachis.
  (**d**) Axillary bud.
  (**e**) Canker starting from a leaf-trimming wound.
  (**f**) Spots on stem.

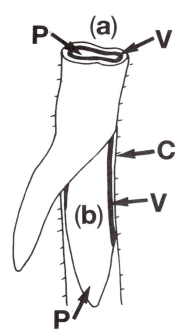

To make these observations, cut the stem longitudinally (and/or transversely) over its entire length

What can you see on the inside of the stem?

182 (**a**) Transverse section.
  (**b**) Longitudinal section.
      P = pitch, V = vascular tissue, C = cortex (skin plus cortical tissues).

183

184

185

186

Examples of changes to the stem (183–189)

183 Cracking of the stem.

184 Cankers.

185 Browning of the stem.

186 Internal necrosis.

187

188

189

187 Beige canker, dry in appearance, on stem—*Botrytis cinerea*.

189 Beige discoloration, partially covered with a white woolly felt—*Sclerotinia sclerotiorum*.

188 Canker with brown edges, lighter in the centre, clearly outlined—*Didymella lycopersici*.

Stem changes and necrosis

(often starting from pruning wounds)

POSSIBLE CAUSES

- *Clavibacter michiganensis* subsp. *michiganensis* (description 3, page 153)
- *Alternaria dauci* f.sp. *solani* (description 6, page 157)
- *Botryosporium* sp.
- *Botrytis cinerea* (description 7, page 158)
- *Didymella lycopersici* (description 19, page 174)
- *Sclerotinia sclerotiorum* (description 22, page 178)
- Hail damage

ADDITIONAL INFORMATION FOR DIAGNOSIS

In order to encourage the development of fungal fructifications and allow their observation, the affected parts of the stem can be placed for several days in a sealed container or plastic bag containing a moist tissue.

	Botrytis cinerea	*Didymella lycopersici*	*Sclerotinia sclerotiorum*
Highly characteristic symptoms	Grey mould on canker	Black spots on canker	Thick white mycelium, large black sclerotia on the canker and in the mould
Symptoms on other parts of the plant	Fruit, leaves, collar	Fruit (rare), leaves (rare)	Collar (rare)
Presence in crop	Very frequent, constant	Infrequent	Infrequent

190 Start of stem disease, dark brown and covered with dark grey mould—*Botrytis cinerea*.

192 Stem canker clearly outlined, covered with numerous black spots (fructifications) —*Didymella lycopersici*.

191 Stem necrosis, beige to light brown, with several large sclerotia, light at first then black —*Sclerotinia sclerotiorum*.

● *Alternaria dauci* f.sp. *solani*

This fungus, parasitic to leaves and fruits, also causes fairly characteristic spots on the stem. These are often clearly outlined, dark brown, and with fine concentric rings (**194**).

193

● *Clavibacter michiganensis* subsp. *michiganensis*
As on the leaves, this bacterium occasionally causes small cankerous white spots on the stem, which then become necrotic in the centre (the same as on the leaves, **193**). It also causes other symptoms—see page 110.

195

● *Botryosporium* sp. (a curiosity)
Quite often, on tissues of tomatoes grown under protection, a saprophytic fungus appears, characterized by numerous white fructifications perpendicular to the substrate (**195**). This colonizes senescent tissues, and does not affect the crop.

● Hail damage
After heavy rain accompanied by hail, it is not uncommon to see localized cracks on the more exposed parts of the stem penduncles (**196**), which correspond to the point of impact of the hailstones. The fruits are often badly affected (**245, 246**).

194

196

197

198

199

200

197 Yellowing, wilting of plants, parts of stem brown—*Erwinia* sp.

198 Pronounced, even browning of stem—*Erwinia* sp.

199 Brown pith, hollow in centre—*Pseudomonas corrugata*.

200 Diffuse browning of the stem—*Pseudomonas corrugata*.

Browning, bronzing of the stem

POSSIBLE CAUSES

- *Erwinia* **sp.** (description 5, page 155)
- *Pseudomonas corrugata* (description 4, page 154)
- *Phytophthora infestans* (see page 65) (description 10, page 162)
- **Various viruses** (descriptions 26 CMV, 29 AMV, pages 184, 186)
- *Aculops lycopersici*
- **Phytotoxicity** (description 31, page 190)

ADDITIONAL INFORMATION FOR DIAGNOSIS

- *Erwinia* **sp.**
- *Pseudomonas corrugata*

Several **bacteria** can cause browning of the stem (sometimes starting from pruning wounds in the case of *Erwinia* spp.) and the pith of the tomato. The most common is *Pseudomonas corrugata*, which sometimes causes the same symptoms but in a weaker form, as well as other symptoms (see also page 116). More rarely in France, *Erwinia* sp. has been associated with similar stem damage. Recent work has shown the possible interaction of two Erwinia: *Erwinia carotovora* subsp. *carotovora* and *Erwinia carotovora* subsp. *atroseptica*, and also sometimes other bacteria such as *Pseudomonas fluorescens* biotype 1 and *Pseudomonas cichorii*.

The conditions facilitating the development of these bacteria, and hence the appearance of symptoms, are the same:
- Plants grown under protection.
- Often very vigorous plants.
- High humidities under protection.
- Presence of free water on the plants.
- Cloudy weather preceding the appearance of the symptoms.

Viruses sometimes cause (in addition to other symptoms, see pages 20, 32, 34, 35, etc.) the appearance of long bronzed or brown areas (in particular alfalfa mosaic virus, **201**—and cucumber mosaic virus, **202**) varying in length from a few centimetres to several decimetres. These can be located at the apex of the plant, on the stem and the petioles. In cases of early infection, entire plants can become necrotic (see page 71).

Aculops lycopersici causes the bronzing of several parts of the tomato (leaves, stems, petioles, etc., but in particular the stem (**203** and see Appendix 1). Attacks on fruits are highly characteristic.

Sometimes brown spots or areas (**204**) are seen on the stem, following the absorption by the roots of herbicides causing **phytotoxicity** (see also pages 23, 41, 49, 54, 71, etc.).

201 Bronze-coloured necrosis, superficial, on one side of the stem only at the apex of a plant—alfalfa mosaic virus.

203 Superficially bronzed stem—*Aculops lycopersici.*

202 Bronze-coloured discoloration, superficial, at the axil of a petiole—cucumber mosaic virus.

204 Drying of leaflets and leaves, brown areas on stem—herbicide phytotoxicity.

Bruising, ruptures, growth of adventitious roots on the stem

It is not uncommon to see the growth of adventitious roots on the stem. These appear progressively. First numerous small, isolated or aligned protuberances appear. These then split, giving rise to a root which may remain very short (preliminary root) or reach several centimetres. In some cases, the roots appear more suddenly, stems break and the cortical tissues break off.

This phenomenon can occur at all levels of the stem, both near the collar and in the middle (very rare at the top). It is often an indicator of poor 'plant functioning'. In fact, when plants are too vigorous or when their supply of water and nutritional elements is disturbed (in the case of changes to the roots or stem), they often produce adventitious roots. This is particularly the case with attacks of pith necrosis. This bacterial disease attacks the pith and the cortical tissues of the stem (see pages 108, 116 and 117—*Pseudomonas corrugata*).

Careful observation, in particular inside the **stems** and **root systems** of the plants affected is therefore recommended.

Certain herbicides absorbed by the roots can also lead to the appearance of many preliminary roots on the stem, in addition to the foliar symptoms.

205

206

207

205 Bruised stem. The bruises are sometimes aligned along the stem.

206 Splitting of stem. Numerous short adventitious roots can be seen in the interior.

207 Peeling, browning of cortical tissues, numerous adventitious roots visible.

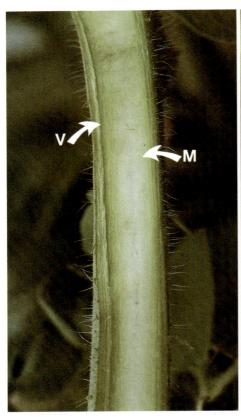

208 Normal stem interior. M = pith, V = vessels

209 Yellowing of pith at edges of vessel—*Clavibacter michiganensis.*

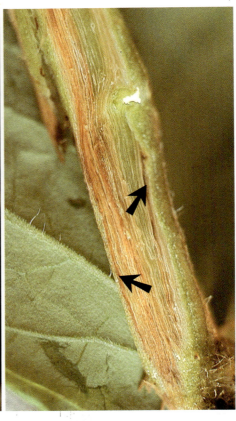

210 Marked browning of vessels which show darker lines in places—*Fusarium oxysporum* f.sp. *lycopersici.*

211 Moist browning of pith and edges of vessels—*Pseudomonas corrugata.*

212 'Dry' browning and hollowing of pith at edges of vessels—*Clavibacter michiganensis.*

213 Browning and liquefaction of the pith —*Pseudomonas corrugata.*

● *Fusarium oxysporum* f.sp. *lycopersici*

Many varieties of tomato are now resistant to Fusarium wilt (see Appendix 2), and this should be taken into account in a diagnosis. If Fusarium wilt is diagnosed although the variety is resistant, several explanations are possible:

- This is indeed Fusarium wilt, but a **new strain** (adapted) has overcome the genetic resistance of the variety (contact a diagnostic laboratory as soon as possible: this type of observation may be of interest to them).
- This is Fusarium wilt but it has developed in the plant because its root system is damaged; in fact, during severe attacks of nematodes or in cases of root asphyxia, the development of *Fusarium oxysporum* f.sp. *lycopersici* in plants which are normally resistant can often be seen.
- Some seeds of a variety sensitive to Fusarium wilt have been mixed with your seeds.
- The diagnosis is wrong. Start observations again and take more care.

In some cases, adventitious roots can appear on the stem (205). If you have the slightest doubt, it is best to take or **send some samples to a specialist laboratory** where the necessary microbiological tests can be carried out.

224

224 Unilateral yellowing of rachis of a leaf.

222 Unilateral yellowing of a leaflet.

222

225

226

227

228

Development of foliar discoloration

225 Yellowing and slight wilting of part of the lamina (in the shape of a V in relation to the petiole).

227 Yellowing and wilting between the main veins of the leaflet.

226 Yellowing at the start of wilting of part of the lamina.

228 Unilateral yellowing of the leaflets of a leaf.

114

- *Verticillium dahliae*
- *Verticillium albo-atrum*

This is a fairly common disease to which many varieties are resistant (see Appendix 2). It mainly attacks in the spring and autumn; in fact, its development in plants is slowed or even inhibited when temperatures exceed 30°C (this sometimes explains the reversible nature of the disease). The vascular browning it causes is less marked than that caused by Fusarium wilt. If you are in any doubt, **consult a specialist laboratory** as confusion is possible.

229

229 Light grey to light brown colouring of vessels (longitudinal section view).

230 Chlorosis and wilting of leaflets (in particular at tops of plants).

232 Glossy pith, darker brown near vessels.

231 Browning of part of the stem and growth of numerous adventitious roots (causing the split).

233 Longitudinal browning on the rachis of a leaf.

- **Pseudomonas corrugata**

This disease mainly attacks very vigorous plants (cultivated under protection) with very large root systems. Often, the growth of the roots is blocked, and they begin to resemble a bottleneck at the ends.

The leaflets turn yellow (230) and wilt, and brown areas appear on the stem and rachis (231, 233). The pith is damaged to some extent (213, 232). Plants only slightly affected may recover. Other diagnostic features are given on page 107.

234

● *Fusarium oxysporum* **f.sp.** *radicis-lycopersici*
This fungus causes browning of the vessels, and can rise to 30 cm above the collar (**234**). It is mainly a **parasite of the roots and collar**. See pages 78, 79, 89 and 90.

5. Irregularities and changes to the fruit

SYMPTOMS STUDIED

- Small spots on fruits
- Extensive spots on fruits
- Soft rotting of fruit
- Changes to the peduncular end of the fruits
- Changes to the stylar end of the fruits
- Presence of rings, circles, etc. on fruits
- Fairly pronounced browning of fruits
- Fruit discoloration
- Fruit deformation
- Cracks, splits of varying sizes on fruits
- Other changes to fruit

POSSIBLE CAUSES

- *Clavibacter michiganensis* subsp. *michiganensis*
- *Pseudomonas syringae* pv. *tomato*
- *Xanthomonas campestris* pv. *vesicatoria*
- *Erwinia* sp.
- *Alternaria dauci* f.sp. *solani*
- *Alternaria tenuis, A. tenuissima*
- *Botrytis cinerea*
- *Colletotrichum coccodes*
- *Fusarium sp., Geotrichum candidum*
- *Mucor* sp., *Rhizopus nigricans*
- *Phytophthora infestans*
- *Phytophthora nicotinae* var. *parasitica*
- *Pleospora herbarum*
- *Rhizoctonia solani*
- Rhabdovirus
- Cucumber mosaic virus
- Alfalfa mosaic virus
- Tobacco mosaic virus
- Various viruses
- Climatic and cultural problems
- Corky peduncular scar (catface)
- Corky stylar scar
- Yellow 'collar'
- Sunburn
- Vibrator damage
- Hail damage
- Bird damage
- Growth splits
- 'Fruit pox'
- 'Internal browning'
- Blotchy ripening
- Blossom end rot

EASY DIAGNOSIS

Most changes observed on the fruits are highly characteristic and their cause can be identified easily.

Certain diseases cause symptoms only on fruits, while others also cause symptoms on other parts of the plant (it is useful to look for these on the plant to confirm the diagnosis).

The most frequent symptom is the production of spots. In order to distinguish between these, various criteria must be used, such as the position on the fruit or the particular symptom (for example, a change to the peduncular scar or browning could be regarded as spots). It is sometimes useful to check several symptoms at the same time if you cannot find the cause of your problem immediately.

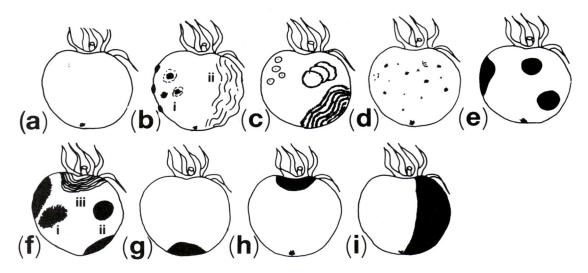

Some aspects and locations of changes to fruits

235   (a) Healthy fruit.
    (b) Spots with haloes (i). Part of fruit bruised (ii).
    (c) Fruit with rings, circles, patches, etc.
    (d) Small spots on fruit.
    (e) Spots on fruit.
    (f) Spots with diffuse edge (i). Spots with well-defined edge (ii). Concentric spots (iii).
    (g) Change localized at the stylar scar (end of fruit).
    (h) Change localized at the peduncular scar (axil of calyx).
    (i) Spot localized on one side of fruit.

236

S = sepal, P = penduncle

P = peduncular end, S = stylar end,
C = columella, T = placental tissues,
PE = pericarp

237

Examples of changes to fruit (238–246)

238 Brown spots on fruit.

239 Browning of a fruit.

240 Discoloration of fruits.

241 Fruit rot.

239

240

241

238

242

243

244

245

246

Small spots on fruits

POSSIBLE CAUSES

- *Clavibacter michiganensis* subsp. *michiganensis*
- *Pseudomonas syringae* pv. *tomato*
- *Xanthomonas campestris* pv. *vesicatoria*
- **Hail damage**

ADDITIONAL INFORMATION FOR DIAGNOSIS

The spots caused by these three bacteria are highly typical, and can be identified with some certainty. When spots are found, there will certainly be other symptoms elsewhere on the plants, in particular on the leaves (*Pseudomonas-Xanthomonas*, see page 48) and on the stems (*Clavibacter*, see pages 108–110). Similarly, if the spots on the fruits are caused by hail, you will also find this symptom on the stem (see page 103).

---

242 Small circular brown fly spots—*Pseudomonas syringae* pv. *tomato*.

244 Superficial corky pustules with an oil halo—*Xanthomonas campestris* pv. *vesicatoria*.

246 Large brown spots 'often open', with chlorosis at the edges—hail damage.

243 Small white spots, brown in centre (like a bird's eye)—*Clavibacter michiganensis* subsp. *michiganensis*.

245 Recessed corky spots, split, brown at edges—hail damage.

247 Small recessed spots, brown in the middle, easily confused with those caused by *Colletotrichum coccodes*—*Alternaria* sp.

248 Large brown spots, slightly recessed, with an irregular coating of black mould—*Alternaria* sp.

249 Well-developed spots, collapsed tissues, with the cuticle of the fruit broken—*Alternaria* sp.

250 *Alternaria* sp. colonizing from splits and blackening adjacent tissues.

Extensive spots on fruits

POSSIBLE CAUSES

- *Alternaria* **sp.** (description 12, page 164)
- *Colletotrichum coccodes* (description 12, page 164)
- *Pleospora herbarum* (description 12, page 164)
- **Sunburn**

ADDITIONAL INFORMATION FOR DIAGNOSIS

Several *Alternaria* spp. are present on the changed fruit, causing symptoms which are often identical: *Alternaria tenuis, Alternaria tenuissima, Alternaria chartarum*, etc. These symptoms are difficult to dissociate from those caused by *Colletotrichum coccodes* (responsible for anthracosis of the tomato) and *Pleospora herbarum* (*Stemphylium botryosum*). These fungi have characteristics in common:

- They mainly colonize the fruits of field tomato crops (in particular unstaked crops).
- The changes they cause are clear and well defined to begin with, and develop slowly. They brown to some extent in the centre, because of the development on or below the cuticle of the mycelium or brown spores of the fungus.
- Although, in some cases, contamination occurs early when the fruit is first forming or when it is still green, the symptoms appear from the turning stage (start of maturity). The riper the fruit, the higher the number of spots on the fruit and the quicker their development.
- They penetrate the fruit either directly through the cuticle or through various wounds (punctures, splits, perforations, micro-lesions on the cuticle, etc.).

Bacteria occur alone or in combination with micro-organisms that are responsible for soft rots (see page 129). They can be regarded as the cause of the bloom on rotted fruit in field crops, where *Alternaria* spp. cause the most damage.

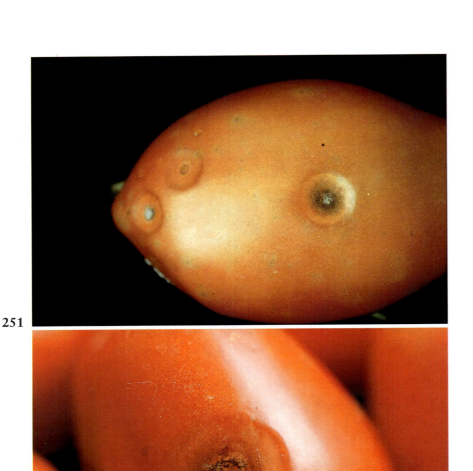

251

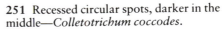

251 Recessed circular spots, darker in the middle—*Colletotrichum coccodes*.

252 More developed spot, concave, with black punctures in the centre—*Colletotrichum coccodes*.

253 Spot covered with numerous superficial black fructifications (fungal spore cushions = acervuli)—*Colletotrichum coccodes*.

254 Large concave spot, darker brown in the centre, covered with numerous concentric black points (fungal spore cases = perithecia) —*Pleospora herbarum*.

252

253

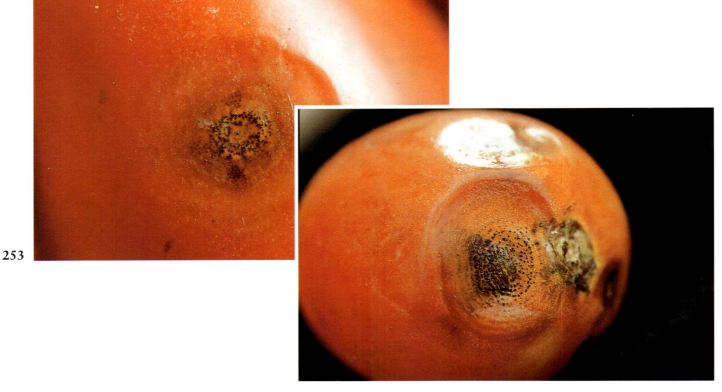

254

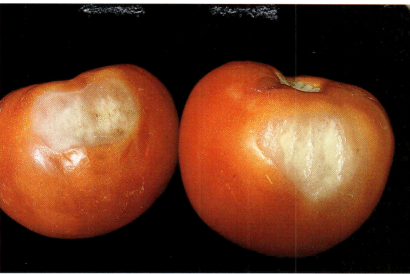

256 White, shiny spots, slightly blistered on one side of fruit—sunburn.

255 White dry patch, concave, on the side of the fruit exposed to the sun—sunburn.

- **Sunburn**

Spots are frequently observed on the side of fruit exposed to the sun; few fruits are usually affected. Sometimes the damage is more serious, in particular when the fruit is less protected from the sun by foliage. The reduction in foliage may have been caused by:

- Severe attacks of air-borne diseases—in particular fungal pathogens (*Alternaria*, early blight; *Stemphylium*, late blight, etc.).
- Excessive leaf fall.
- Turning of plants at the first harvest of field-grown unstaked crops.

This non-parasitic disease often affects green plants in the field crops.

257 Alterations caused by several rot agents on the fruits—micro-organism complex.

258 Very dense felting consisting of numerous, minute black pinheads—*Rhizopus nigricans*.

259 'White foam' covering the flesh of a cracked fruit—*Geotrichum candidum*.

260 Very dense grey felting, covered with numerous grey pinheads—*Mucor* sp.

Soft rots of the fruits

POSSIBLE CAUSES

- *Erwinia* **sp.** (descriptions 5–12, pages 155–164)
- *Fusarium* **sp.** (description 12, page 164)
- *Geotrichum candidum* (description 12, page 164)
- *Mucor* **sp.** (description 12, page 164)
- *Rhizopus nigricans* (description 12, page 164)

ADDITIONAL INFORMATION FOR DIAGNOSIS

Numerous micro-organisms, alone or in combination, can cause changes that rapidly become soft and 'wet', and which lead to the total collapse of the flesh of the fruits. These include several bacteria (*Erwinia*, in particular *Erwinia carotovora*, and *Bacillus* spp.), *Botrytis cinerea*, several species of *Fusarium*, *Geotrichum candidum*, *Pullularia pullulans*, *Pythium*, Mucorales (*Rhizopus nigricans* and various *Mucor* spp., including *Mucor hiemalis*). All these micro-organisms have common characteristics:

- They cause very similar changes, which often develop quickly, accompanied by liquefaction and collapse of the internal tissues. The skin folds and cracks. Only the fungal spores or the bacterial ooze on the surface of the fruits allow identification in some cases.

- They mainly affect ripe fruit, more particularly over-ripe fruit of unstaked plants in the field (much more frequently than staked plants).

- They penetrate fruit via various wounds (insect punctures, splits, perforations, micro-lesions of the cuticles, etc.).

- They are strongly influenced by the agricultural and climatic conditions affecting the crop. For example, poor irrigation or relatively heavy rain after a period of drought can cause fruit to burst, from which rot can develop quickly.

The micro-organisms most often found are:

- *Fusarium* **sp.** (white felting of the mycelium, cotton-like on the affected area).
- *Geotrichum candidum* (white foam on the affected area).
- *Rhizopus nigricans* (dense felting with numerous black pinheads on the affected area).
- *Erwinia* **sp.** (creamy white ooze on the affected area).

All these micro-organisms occur either alone or in association, in particular with *Alternaria*, which also causes symptoms on the fruits (see illustrations of extensive spots on fruits on page 125). They can be regarded as the cause of the bloom on rotted fruit in field crops, where *Rhizopius* and *Mucor* cause the most damage.

261

262

263

264

261 Dark brown spot, recessed, divided into zones, with necrotic sepals—*Alternaria dauci* f.sp. *solani*.

262 Soft rot, grey to beige, quickly covered with a grey mould—*Botrytis cinerea*.

263 Collapse of tissues at the peduncular end of the fruit, which is covered by a black mould—*Alternaria tenuis*.

264 Corky area around the peduncular scar—corky peduncular scar.

Changes to the peduncular end of the fruits

POSSIBLE CAUSES

- *Alternaria dauci* **f. sp.** *solani* (description 6, page 157)
- *Alternaria* **spp.** (description 12, page 164)
- *Botrytis cinerea* (description 7, page 158)
- **Corky peduncular scar**

ADDITIONAL INFORMATION FOR DIAGNOSIS

The symptoms on fruit caused by the two parasitic fungi *(Alternaria dauci* f.sp. *solani* and *Botrytis cinerea* are highly characteristic and can be readily identified. At the same time, they also cause other symptoms on the plant (see pages 60–63).

Alternaria tenuis and other *Alternaria* spp. can, under some conditions, cause fruit rot (mainly when over-ripe) beginning from senescent sepals colonized by these bacteria (see page 125, where the other fungi are studied in detail).

The large-fruited varieties of tomato (multilocular with few seeds per loculus and large central columella) often have large **corky peduncular scars**; these must be regarded as a fault associated with these types of variety. Fruits with this symptom often also have a pronounced corky stylar scar.

265

266

267

268

265 Small necrotic spots, light brown at the edge of the fruit—blossom end rot.

267 Black mould (spores of a saprophytic *Alternaria* spp.) colonizing blossom end rot—blossom end rot.

266 Patches with black spots, concave at the end of the fruits, sometimes penetrating deeply—blossom end rot.

268 Large corky scars and craters at the ends of the fruits—corky stylar scar (catface).

Changes to the stylar end of the fruits

POSSIBLE CAUSES

- **Corky stylar scar** (catface)
- **Blossom end rot** ('black end')

ADDITIONAL INFORMATION FOR DIAGNOSIS

- **Corky stylar scar**: this non-parasitic disease is fairly common. It mainly affects fruits (especially the multilocular varieties) of early crops (under protection and in the field) which are subjected to unfavourable climatic conditions (temperatures too low) during blossoming and fruit formation (see page 143). Sometimes the fruit can be very deformed (see pages 142–143). Similar damage can be caused by growth regulator type herbicides such as 2-4 D.

- **Blossom end rot**: this physiological disorder is also often observed in all types of crops, more particularly in crops with furrow or flood irrigation. The appearance of this symptom appears to bè closely linked to:

- A lack of calcium in the fruit (caused by real or induced calcium deficiency).

- The irrigation technique used: the attacks are more severe in crops with furrow irrigation compared to drip irrigation. In addition to water shortages which can occur between applications (excessive interval), infrequent large quantities of water can result in temporary asphyxiation, reducing root absorption and leading to a deficiency (a motto to follow: water little and often rather than too much).

- Under certain agricultural conditions (excess salinity, excess nitrogen, poor soil preparation, soil working 'mutilating' roots, etc.).

- Attacks of soil-borne parasites affecting the roots.

- Certain varieties which are more sensitive than others.

269

Symptoms on fruits not necessarily in contact with the ground

269 White, clearly defined rings with a small brown puncture in the middle—*Botrytis cinerea*.

270 Yellowish white rings with diffuse edges, on ripe fruit—*Botrytis cinerea*.

271 Numerous 'circles' and spots covering the entire surface of a brown, slightly bruised fruit—alfalfa mosaic virus.

272 Whitish yellow rings and arcs on a fruit (blistered appearance)—cucumber mosaic virus.

270

271

272

273

274

275

POSSIBLE CAUSES

- *Botrytis cinerea* (description 7, page 158)
- *Phytophthora nicotinae* var. *parasitica* (description 20, page 175)
- *Rhizoctonia solani* (description 21, page 177)
- **Various viruses**

(Description 26, page 184—CMV; description 28, page 186—AMV)

ADDITIONAL INFORMATION FOR DIAGNOSIS

These symptoms on fruit are normally characteristic enough (except in the case of viruses) for easy diagnosis. If you suspect a viral disease, consult a specialist laboratory.

Symptoms developing in particular on parts of fruit in contact with the ground

273 Brown semi-circles, concentric, with looped edges—*Phytophthora nicotinae* var. *parasitica*.

274 Light brown to brown spot, slightly corky in the middle, with concentric circles, numerous small cankerous spots around—*Rhizoctonia solani*.

275 Various changes caused by *Rhizoctonia solani*: concentric spots may crack. Sometimes accompanied by a brown area, numerous brown punctures, superficial brown lines (= fungus mycelium)—*Rhizoctonia solani*.

280

281

282

283

284

Air-borne and soil-borne blights

	Viruses	Blights	
		Late blight *Phytophthora infestans*	Buck-eye rot *Phytophthora parasitica*
Crops affected:			
• Field, unstaked	+	+	+
• Protected	+	+	+/−
Presence of other symptoms on plant	+/−	++	−
Fruits affected mainly in contact with ground	−	−	+
Numerous fruits affected, disease spread through plant	+	−	−
Browning in concentric circles	+/−	−	++

280 Part of fruit brown, mottled and with irregular bruising, with poorly defined edge—*Phytophthora infestans.*

282 Brown spot, lighter at edges—*Phytophthora infestans.*

283 Large brown spots consisting of concentric rings with looped edges (typical buck-eye rot symptoms)—*Phytophthora nicotinae* var. *parasitica.*

281 Spots initially brown on several fruits of a truss—*Phytophthora nicotinae* var. *parasitica.*

284 Brown spot developing round the peduncular end, numerous pale brown spots on leaves—*Phytophthora infestans.*

285

286

287

285 Mottling on ripe fruit—Rhabdovirus (see page 25).

286 Peduncular end of fruit remaining yellow despite state of advanced maturity—yellow neck.

287 Ripe fruits mottled with green, diffuse areas—blotchy ripening or tobacco mosaic virus.

288 Tissues beneath skin and 'vessels' brown and corky (internal browning)—blotchy ripening or tobacco mosaic virus.

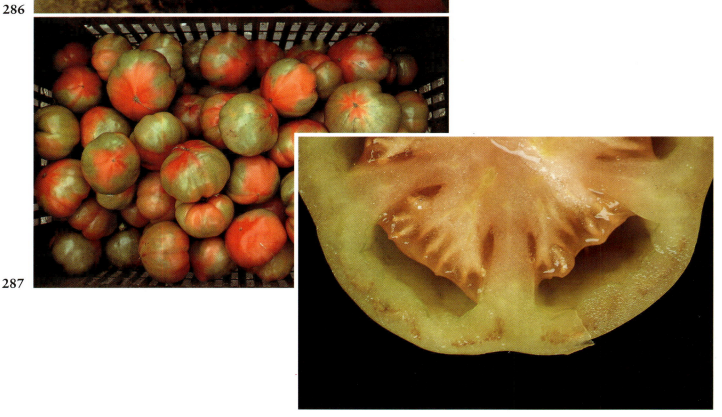

288

Fruit discoloration

POSSIBLE CAUSES

- **Rhabdovirus and various other viruses**
- **Tobacco mosaic virus (TMV)** (description 25, page 183)
- **Yellow neck** (description 31, page 190)
- **Blotchy ripening** (description 31, page 190)

ADDITIONAL INFORMATION FOR DIAGNOSIS

- **Various viruses** can cause mosaics, mottling and discoloured patches (of various shapes) on the fruit: these cannot be identified from these symptoms alone. If there are symptoms on other parts of the plant, see the sections concerned or consult a specialist laboratory (which is preferable, given the difficulty of diagnosing viruses).

- **Tobacco mosaic virus** is now fairly rare as many varieties are resistant (see the list in Appendix 2). Brown necrotic patches are sometimes found on the fruit of resistant varieties (**278, 279**) as a result of a break in the resistance (especially when resistant varieties are placed near affected sensitive plants, when temperatures are high and light intense). The affected plants sometimes have symptoms on the leaflets (see pages 21–33).

Both tobacco mosaic virus and a non-parasitic disease (blotchy ripening of tomatoes) are responsible for blotchy ripening of fruits and internal browning which may be localized in cells surrounding vascular bundles in the pericarp.

- **Blotchy ripening** is a non-parasitic disease, still not well known, which has the following characteristics:
  - There are differences in the sensitivity of varieties.
  - Deficiencies of potassium and boron can be linked to its appearance.
  - It mainly occurs in periods of reduced light with high temperatures.
  - Low levels of soluble salts in the soil or in the nutrient solution will encourage it.

In general, it mainly affects protected crops particularly in early spring or late autumn.

It is frequently associated with symptoms of internal browning and grey wall, which also appear to be induced by climatic and cultural conditions similar to those that encourage blotchy ripening. Many authors consider that all three symptoms are part of the same disease complex.

- **Yellow neck** is also a non-parasitic disease, which appears to occur under the following conditions:
  - Temperatures above 25°C when fruit is ripening.
  - Major temperature differences between day and night.
  - Low temperatures during the growth of the fruit.
  - Temperatures above 22°C after picking.
  - Plants with reduced foliage or after excessive leaf fall.
  - Unbalanced use of fertilizers, particularly low in potassium.
  - Certain varieties are more sensitive.

289

289 Fruit heavily ribbed and deformed.

29

290 Fruit hollow and deformed.

292 Highly deformed fruit with corky scars of varying sizes, cracked (see page 133)—corky scars (catface).

291

291 Stylar end of fruit pointed. Teated fruits.

29

293

293 Numerous bumps at end of fruit—bumpy fruit.

Fruit deformations

POSSIBLE CAUSES

- **Corky stylar scar** (catface)
- **Poor cultural and climatic conditions**
- **Various phytotoxicities** (description 31, page 190)

ADDITIONAL INFORMATION FOR DIAGNOSIS

The current trend in cultural schedules is to plant crops as early as possible, at times of the year when the climate is not very favourable for flowering nor in particular for fruit set.

Several **cultural and climatic factors** can disrupt the formation of the flowers and pollen, and the release of the pollen:

- Ground too cold, leading to a potassium deficiency.
- Daylight periods too short and with light intensity too low.
- Excess nitrogen.
- Relatively high humidity or atmosphere too dry.
- Ambient temperatures too low, in particular at night, or too high.

Under these conditions, the crops (both under protection with inadequate heating and in the field) give deformed fruits on the first truss.

In order to remedy these poor conditions, the use of fruit-setting substances is standard practice which, when poorly applied, cause some risk of **phytotoxicity**.

- Deformation and reduction in surface area of leaves (see page 22).
- Heavily ribbed fruit, hollow, treated, deformed or soft (**289–293**).

If these symptoms are found, the following questions should be asked:

- Has the specified dose of fruit-setting product been used?
- Has the treatement been repeated on the same flowers?
- Has product powder been sprinkled on to the leaves and terminal buds?
- Was the product applied in poor conditions (temperatures too low)?

294

295

294 Radial split localized in the peduncular area of the fruit—growth splits.

295 Radial cracks starting at the side of the stylar scar—growth splits.

296 Small cankerous spots and radial splits on green fruit—*Clavibacter michiganensis* subsp. *michiganensis*.

297 Small splits with 'clear dry edges'—fruit pox.

296

297

Cracks, splits of varying sizes on fruits

POSSIBLE CAUSES

- *Clavibacter michiganensis* subsp. *michiganensis* (descriptions 3, page 153)
- **Growth splits**
- **Fruit pox**
- **Vibrator damage**
- **Bird damage**
- **Hail damage** (see pages 122–123)
- **'Russeting' phytotoxicities**

ADDITIONAL INFORMATION FOR DIAGNOSIS

- **Growth splits** are non-parasitic in origin. They appear in various situations (more particularly in staked crops in the field):
  - When a period of humidity follows a period of drought, for example, rain or major irrigation in crops which have lacked water. This applies in particular to furrow irrigated crops.
  - When watered irregularly.
  - When a nitrogen fertilizer is scattered.
  - After a sudden increase in temperature leading to a rapid increase in growth of the fruits.

Fruits in these conditions appear unable to 'absorb' the excessive water entering them: they burst with radial or concentric splits.

Weaker plants with reduced foliage are more sensitive. Their low rate of evapo-transpiration does not balance water uptake adequately, resulting in an excess of water reaching the fruit.

There are pronounced differences between varieties, for example large-fruited varieties are more sensitive to this disorder.

- *Clavibacter michiganensis* subsp. *michiganensis* can, in rare cases, cause cankers and small splits on the fruits. These are always accompanied by other symptoms on the plants (see pages 108 and 110).

- **Fruit pox or tomato pox** is a genetic disorder, affecting the fruit of tomato varieties in which a recessive gene appears. It is rarely seen in France and remains a minor problem. It mainly occurs on green fruit in the form of small spots of darker green. When the fruits ripen, they become lighter, crack and dry out. The conditions predisposing to these disorders are not known.

298

299

300

There may be many other causes of a rupture of the cuticle and skin of fruits. In particular, these include:

- **Vibrator damage**: when the flower trusses are shaken, the operator may sometimes catch the fruit with his tool (vibrator). Such damage causes wounds of varying depths (**298**) which often appear as corky scars. Other accidental damage can cause identical symptoms.

- **Russeting phytotoxicities**: when certain pesticides are applied in excessive quantities or under poor conditions, the superficial browning sometimes caused on the fruit makes their cuticles more corky and hence less elastic, so during growth of the fruits, they tend to develop a number of small cracks which are often concentric (**299**).

The same (especially the corky appearance of the cuticle) or similar symptoms (which the Dutch call russeting) occur frequently in heated greenhouses. The main factors predisposing to their development are:

- Daytime temperatures too high (accompanied by relatively low humidities) and night temperatures too low.
- Average temperatures too low in the greenhouses.
- Periods of cloudy weather, poor light intensity in greenhouses.
- Presence of a thermal screen.
- Low soluble salt level in the nutrient solutions.
- Different sensitivities of varieties.

- **Birds**: ripe tomatoes are appetizing meals for many species of bird, which can cause major damage by pecking (**300**).

146

Other changes to fruit

● Bug damage

Several species of bug can cause minute spots on the tomato. On green fruit, these look like pin pricks around which the tissues are lighter than in the rest of the fruit. On ripe fruit, the spots are white to yellow (**301**) and correspond to the pricks made by the insects for feeding. During feeding, enzymes are released which cause the white and spongy appearance (cloudy spot) of the tissues of the pericarp just below the punctures. These areas of damaged tissue (firm to the touch) are transparent and appear as yellow spots with diffuse edges.

● Unidentified non-parasitic problem

Browning of the internal tissues of the fruit (placental tissues, columella) (**302**) has frequently been reported. The cause of this is unknown. The only information available is as follows:

- no micro-organism has been isolated. It appears that this is a non-parasitic disease.

- The problem mainly occurs in fruits of protected crops.

- There appear to be differences in sensitivity between varieties.

- The symptoms appear inside both green fruit and ripe fruit, often following periods of hot weather with very high temperatures in the greenhouse.

  Similar symptoms have been observed in tomatoes from early crops under protection (in North Africa). The variety used was not perfectly adapted to the production period, and cultural errors appear to have been made.

301

302

303

● **Damage caused by poor climatic conditions**

Fruits from field tomato crops (mainly staked) in autumn and winter (Spain, etc.) sometimes show various changes mainly localized in the upper part of the fruit around the peduncle. Most common is the appearance of brown spots or patches, with diffuse edges, superficial or recessed to some degree (**303**). These are caused by the climatic conditions (nights relatively cold and humid) which occur at these times of the year, particularly the autumn. The cuticle and skin tissues are more or less affected by low temperatures at the surface of the fruits and by the free water present. These damaged areas are then colonized by various micro-organisms, in particular by *Alternaria* spp. (see page 125).

Part Two
Principal Characteristics of Pathogenic Agents and Methods of Control

6. Bacteria

DESCRIPTION 1

Pseudomonas syringae pv. *tomato* (Okabe) Alstatt
Common name: speckles or specks

SYMPTOMS

See **69, 75, 78, 79, 238, 242.**

PRINCIPAL CHARACTERISTICS

Sources: on seeds, in the soil either directly or via vegetable debris. The bacteria appear able to survive on the roots and foliage of several cultivated plants and weeds.

Dissemination: through rain, overhead irrigation and wind, which can carry drops of water (containing the bacteria) across considerable distances.

Penetration: through stomates and wounds.

Conditions for development: relatively low temperatures, ideally around 20°C; high humidity (dew, fog, rain, spray irrigation), in particular the presence of a film of water on the plants. These conditions, if maintained for 24 hours, are sufficient to ensure development of the disease which appears 8 – 10 days after contamination.

CONTROL METHODS

During cultivation: there is no real cure. It is best to avoid excessive humidity, in particular the presence of free water on the plants. To achieve this, provide maximum ventilation for crops under protection, and avoid overhead irrigation. If this is not possible, water in the mornings (never in the evenings) so that the foliage can dry quickly during the day.

Apply copper as Bordeaux mixture, wetting the plants well, which will restrict the development of this disease. Recent studies have shown that copper mixed just before application with dithiocarbamate fungicide (Mancozeb) appears more effective.

During or after cultivation, remove crop debris.

Next crop: rotate the land, use disinfected seeds (treated with sodium chlorite or chlorinated lime), treat with copper only or in association with Mancozeb from propagation onwards, wetting plants well.

Some recent varieties intended for field culture as unstaked crops are resistant to *P. tomato* (see Appendix 2).

DESCRIPTION 2

Xanthomonas campestris pv. *vesicatoria* (Doidge) Dowson
Causes bacterial scale

SYMPTOMS

See **73, 74, 76, 77, 244.**

PRINCIPAL CHARACTERISTICS

Sources: on seeds and plant debris (on and in the soil).

Conditions for development: this bacterium mainly appears in the summer, and is encouraged by fairly high temperatures (ideally 25°C) and high humidities (storm rains, spraying, etc.).

Other characteristics: the same as those of *Pseudomonas tomato*; these two bacteria often appear at the same time in crops.

CONTROL METHODS

Use the same methods as those described for *P. tomato* (see description 1).

Clavibacter michiganensis subsp. *michiganensis* David et. al = *Corynebacterium michiganense* (E.F.Sm) Jensen
causes bacterial canker

SYMPTOMS

See 93, 124, 126, 193, 209, 212, 214, 215, 216, 217, 243, 296.

PRINCIPAL CHARACTERISTICS

Sources: in the soil (one or more years, depending on authority), plant debris, on various materials (pots, drippers, etc.), on greenhouse structures, tools, seeds.

Dissemination: through rain, overhead irrigation, nutrient solutions in soil-less crops and above all during cultural operations such as pruning and leaf removal (hence the often characteristic distribution in a line in the crop).

Penetration: through wounds such as pruning cuts, roots, stomates.

Conditions for development: 18–24°C with over 80% humidity. Like most bacteria, it is encouraged by humid weather. Very vigorous plants are more sensitive, especially when this results from the excessive application of nitrogen.

CONTROL METHODS

During cultivation: for limited areas, remove and burn the diseased plants and their immediate neighbours which appear healthy. For larger areas, put these into quarantine and treat them last, after the healthy parts of the crop.

Disinfect pruning equipment at regular intervals (at the start of a line) using alcohol or a bleach/disinfectant solution, wash hands frequently with soap and water.

Avoid overhead irrigation. At the end of the crop, remove all debris.

Next crop: at present there are no resistant varieties available (selection programmes are in progress). Use seeds which have given a negative response to the immunofluorescence test (marked on the pack).

The soil or greenhouse area used for propagation must be disinfected; avoid excess humidity and all practices leading to wilting of the plants; carry out copper treatment in the form of Bordeaux mixture from the single leaf stage (200 to 300 g copper metal/hl).

During propagation, avoid excessive manuring and high density sowing, and maintain environmental conditions unfavourable to the pathogen.

Before placing plants under protection, take care to treat the ground with a fumigant such as methyl bromide. All equipment used or re-used during cultivation, such as pots, stakes, string, etc., should be treated with a bleach/disinfectant solution (12% available chlorine, soak for 24 hours in a 2% preparation and rinse with water) or formaldehyde (2–5% formalin, soak 1 hour and cover with a plastic sheet for 24 hours).

DESCRIPTION 4

Pseudomonas corrugata Roberts and Scarlett

Causes pith or pith neurosis

SYMPTOMS

See **183, 186, 199, 200, 205, 206, 207, 211, 213, 230, 231, 232, 233.**

PRINCIPAL CHARACTERISTICS

Black pith is a little known disease which mainly affects crops under protection, and exceptionally field crops in cloudy, humid weather. It appears to be encouraged by excessive nitrogen: diseased plants are often very vigorous (strongly vegetative with fat stems). The disease often appears after periods of cloudy weather in which the humidity levels under the protection have been high.

CONTROL METHODS

During cultivation: unfortunately there is no effective treatment for this disease. It is recommended that over-vigorous plants are avoided by controlling fertilizer use (a fairly high potassium content in the soil or nutrient solution will help).

Prevent high humidity levels under protection by providing maximum ventilation.

Do not forget that this disease is 'reversible': plants partially affected and which have stopped growing can start again and achieve a very satisfactory growth and production as soon as conditions become suitable (warmth and light).

During or after cultivation, remove badly affected or dead plants.

Next crop: repeat the measures described above, but as preventive treatment.

DESCRIPTION 5

Other bacteria attacking the tomato

Several species of **Erwinia**, in particular *Erwinia carotovora*, are likely to cause damage to stems (**185, 197, 198**) and fruits of the tomato as a result of the secretion of numerous cellulolytic and pectinolytic enzymes. These are highly polyphagous bacteria which occur in conditions of high humidity and at temperatures varying between 5°C and 37°C, ideally 22°C. They survive in the soil, on plant debris of diseased plants.

When they attack crops under protection (more rarely in open field), the method of treatment is the same as that for *Pseudomonas corrugata* (see description 4).

Galls caused by **Agrobacterium** sp. are very rarely seen on tomato roots. However, in experiments the tomato has been shown to be highly susceptible. This bacterium survives in the soil as a saprophyte. It is transmitted by water, on tools frequently in contact with the soil, etc.

During cultivation, no method of treatment will effectively control this disease. Affected plants should therefore be destroyed, and contaminated substrates of soil-less crops should be removed.

7. *Fungi*
Fungi attacking the leaves

DESCRIPTION 6

Alternaria dauci f. sp. *solani* (E11. and Mart.) Neerg

Causes alternariosis, target spot, early blight

SYMPTOMS

See 70, 107, 108, 111, 112, 115, 116, 194, 261.

PRINCIPAL CHARACTERISTICS

Sources: in the soil, on vegetable debris.

Dissemination: by wind and rain.

Transmission: by seeds.

Conditions encouraging development: high humidities and temperatures between 18°C and 25°C. Dews or slight showers (5 mm) will encourage its spread, but these must be repeated in order for the disease to develop rapidly. Poorly manured plants or plants laden with fruit are more sensitive.

CONTROL METHODS

During cultivation: to stop development of the disease, treat with anti-mildew contact fungicides* and repeat applications after heavy rain and overhead irrigation; avoid watering plants overhead, especially in the evening, and remove plant debris after cultivation.

Next crop: *Alternaria solani* sometimes causes damage to young plants (stem cankers), so seeds should be treated with thiram or iprodione if they have not been 'treated'. If the plants are propagated in the same place as the previous year, watch carefully as the disease may have occurred here the year before. For greater safety, disinfect the soil.

IMPORTANT NOTE

* Approval of fungicides may vary in different countries and local labels should be consulted before use.

DESCRIPTION 7

Botrytis cinerea Pers.

Causes grey mould

SYMPTOMS

See 105, 106, 109, 110, 113, 114, 163, 172, 173, 184, 187, 190, 262, 269, 270.

PRINCIPAL CHARACTERISTICS

A ubiquitous, polyphagous fungus capable of attacking and colonizing many plants (particularly via wounds, and senescent tissues which form ideal nutritional 'bases' for its development). Amongst these, lettuce, peppers and aubergines are rotated with the tomato or often grown nearby.

Sources: on plant debris and in the soil, in several forms: conidia, mycelium, sclerotia.

Dissemination: by rain, wind and air currents in shelters.

Conditions encouraging development: relative humidity 95%, temperature 17–23°C.

CONTROL METHODS

During cultivation: treatment for this disease is difficult at present as there are many strains of *Botrytis cinerea* which are resistant to the main fungicides used for control. Also, the climatic conditions of crops under protection are perfectly suited for its development. In order to reduce its frequency, certain measures must be taken.

Ventilate the greenhouses as much as possible in order to reduce the humidity; installation of a thermal screen will increase humidity and reduce light intensity. Avoid the presence of free water on the plants wherever possible. Sometimes, if necessary, defoliating the plants will allow better ventilation.

Remove side shoots early and close to the stem in order to keep the wounds as small as possible and hence reduce risk. In the case of stem cankers, these must be painted with a thick fungicide mixture (thiram and/or iprodione mixture). The addition of paraffin oil to this mixture will make it more effective.

Treat plants alternately with fungicides from two different groups:

- **Group 1**: chlorothalonil, dichlofluanide, thiram.
  Botrytis cinerea has no resistance to these traditional broad-spectrum products, but their effectiveness is sometimes limited.

- **Group 2**: benzimidazoles (benomyl, carbendazim), thiophanates, dicarboximides = cyclic imides (iprodione, procimidone, vinclozolin).
  The first two were effective a few years ago but the development of resistant strains (persistent and highly resistant) has made them ineffective. These should no longer be used. Dicarboximides, highly effective in the initial period of their use, have also generated resistant strains (poor resistance, non-persistent) which are now very common in many greenhouses. Dicarboximides should be abandoned for a while, or their use considerably reduced by alternating or replacing with products in the first group.

A combination of fungicides* (vinclozolin + thiram) recently marketed but not approved for use on the tomato could give more satisfactory results. Another combination of fungicides* (diethofencarbe + carbendazim), also not yet approved, could be used for specific treatment of strains resistant to the benzimimazoles and thiophanates (diethofencarbe has a negative crossed resistance with these families). These two fungicides* are currently used with some success on vines, but their use on tomatoes depends on future approval.

IMPORTANT NOTE

* Approval of fungicides may vary in different countries and local labels should be consulted before use.

Remove plant debris from the crop as

soon as possible, particularly damaged plants and fruits.

Avoid vigorous plants.

Next crop: in order to avoid attacks on young plants (from senescent cotyledons or at the collar), do not bury the plants too deeply during pricking out, or earth them up too much when planting. If damage occurs despite these precautions, apply an anti-botrytis fungicide locally (directed spray).

As soon as the leaf canopy becomes dense, take care, particularly during periods of cloudy weather and as harvest approaches.

DESCRIPTION 8

Fulvia fulva (Cke) Ciferri = *Cladosporium fulvum* Cooke

Causes cladosporiosis or leaf mould

SYMPTOMS

See **96, 97, 100, 101**.

PRINCIPAL CHARACTERISTICS

Sources: on plant debris in the soil, on greenhouse structures. The conidia are viable for 9–12 months.

Dissemination: by wind, or air currents in greenhouses.

Penetration in the plant: by stomatas.

Conditions encouraging development: temperatures of the order of 20–25°C and relative humidity above 80%. Contamination takes 24–48 hours in poorly ventilated greenhouses. The incubation period is fairly long—usually 12–15 days, then sporulation occurs within some hours.

CONTROL METHOD

During cultivation: ventilate the greenhouse as much as possible to reduce the relative humidity. Defoliation at the base of the plant will remove the leaves attacked first and increase ventilation of the lower parts of the plant.

Avoid sprinkling and overhead irrigation, particularly in the evening.

Apply treatment with one of the following products* either as a preventive, or as soon as the first spots appear: maneb, Mancozeb, chlorothalonil, triforine, fenarimol. Benzimidazoles (benzomyl, carbendazim, etc.) are also reported to be effective, particularly when applied to the roots (drip irrigation, application to the base of the plants).

Remove plant debris during and after cultivation.

One biological treatment using a natural hyperparasite *Hansfordia pulvinata* (**103**) is currently being developed and may be available for use soon.

Next crop: there are varieties resistant to all or most of the strains present in France (see list in Appendix 2).

Before planting the crop, disinfect the structures and walls of the greenhouse in order to destroy the spores or other propagules of *Fulvia fulva* or other fungi present. To do this, use 2% formaldehyde, sprayed at high pressure against the walls. This product is currently the most popular. It is also used in fumigation at 0.9l of commercial solution (38% formaldehyde) per 100 m³. Potassium permanganate (at the rate of 360 g for this quantity of formaldehyde) is often added as an oxidizing agent. During the application, the temperature should be at least 10°C and the relative humidity should lie between 50 and 80% (avoid moisture on the walls). Leave the greenhouse closed for at least 24 hours and ventilate well for at least one day before planting. Formaldehyde is sometimes used for drenching at the rate of 3500 l/hectare; it is often sprayed in a diluted solution (2 – 5% commercial solution at 30%).

Bleach/disinfectant (or sodium chlorite) is also increasingly used for this type of disinfection, by spraying solutions of 4 to 7% (commercial solution with 48°Chl).

Other products (based on quaternary ammonium or phenolic derivatives) are recommended for the soil and to disinfect the structures and walls of greenhouses.

IMPORTANT NOTE

* Approval of fungicides may vary in different countries and local labels should be consulted before use.

DESCRIPTION 9

Leveillula taurica (Lèv.) Arn. = *Oïdiopsis taurica* (Lèv.) Salm.

Causes oïdium or powdery mildew

SYMPTOMS

See **71, 94, 95, 98, 99, 100**.

PRINCIPAL CHARACTERISTICS

Sources: the fungus colonizes many cultivated plants (Solanaceae, artichokes, cardoons, leeks, etc.) and weeds.

Dissemination: by wind, sometimes over long distances. The spores germinate on the surface of the leaves and the mycelium colonizes the leaf tissue.

Optimum conditions for development: relative humidity 50–70%, temperature 20–25°C.

CONTROL CONDITIONS

During cultivation and next crop: some years ago, preventive chemical treatment was not necessary. Now the disease attacks often and seriously and it is not easy to arrest its development. The disease occurs in France after July, when preventive treatment with wettable sulphur may be carried out. The secondary, acaricidal effects of this treatment are useful at a time when these pests are particularly active.

The development of *Erysiphe* sp. (another powdery mildew), reported only once in France under protection (**102**), was stopped by application of traditional powdery mildew fungicides. At present, no information on this fungus is available.

DESCRIPTION 10

Phytophthora infestans (Mont.) de By

Causes mildew or late blight

SYMPTOMS

See 72, 119, 120, 121, 122, 280, 282, 284.

PRINCIPAL CHARACTERISTICS

Sources: in the soil, from diseased potato tubers. When they germinate, they develop diseased leaves.

Dissemination: by wind and rain, sometimes over long distances.

Penetration: by stomata.

Conditions encouraging development (which is often rapid): high relative humidity greater than 90%, temperatures between 10°C and 25°C. Cold nights and moderately warm days with high humidity encourage the spread of this pathogen, whereas a dry atmosphere and temperatures close to 30°C will destroy it.

Some strains are particularly adapted to the potato, whereas other strains will attack other plants in the Solanaceae.

CONTROL METHODS

During cultivation: avoid overhead irrigation, in particular at the end of the day, and ventilate greenhouses as much as possible.

Many commercial fungicides* can be used as preventive or curative treatments; in the latter case, systemic products are most effective.

IMPORTANT NOTE

* Approval of fungicides may vary in different countries and local labels should be consulted before use.

Contact fungicides.

Systemic fungicides (often associated with one or more contact fungicides).

Remove plant debris during and after cultivation.

Next crop: some field varieties of tomato have partial resistance (see Appendix 2).

Avoid planting tomatoes next to potato crops.

Note warnings of disease development and carry out preventive treatment as soon as the disease is reported in the region.

DESCRIPTION 11

Stemphylium spp.

Causes stemphyliosis or stemphylium leaf spot

Five species can cause this disease:

- *Stemphylium solani* **Weber**
- *Stemphylium floridanum* **Hannon and Weber** = *Stemphylium lycopersici*
- *Stemphylium botryosum* **Wallr.**
- *Stemphylium botryosum* **f. sp.** *lycopersici* **Wallr. (Rotem, Cohen and Wallr.)**
- *Stemphylium vesicarium* **(Wallr.) Simmons**

The latter species is encountered most frequently in the Mediterranean basin, in particular under plastic greenhouses.

SYMPTOMS

See **83** and **85**.

PRINCIPAL CHARACTERISTICS

Sources: in the soil, on plant debris.

Dissemination: by currents of air, wind and rain.

This disease is encouraged by warm and humid weather, and develops quickly in rainy weather but also when morning dews occur.

CONTROL METHODS

All methods used against Alternaria can be used (see description 6). Many varieties are resistant (see Appendix 2).

Fungi responsible for fruit rot

DESCRIPTION 12

Alternaria chartarum **Preuss**
Alternaria tenuis **Auct.** *(Alternaria alternata)*
Alternaria tenuissima **(Fr.) Wilts.**
Botrytis cinerea **Pers. (=** *Sclerotinia fuckeliana* **(de By.) Fuck.)**
Cladosporium herbarum **(Pers.) L.K.**
Colletotrichum coccodes **(Wallr.) Hugues. =** *C. atramentarium* **(Berk. and Br.) Taub.**
Fusarium **sp.**
Geotrichum candidum **Link =** *Oospora lactis* **(Fr.) Lind.**
Helminthosporium **sp.**
Phoma **sp.**
Pullularia pullulans **(de Bary) Kerkh =** *Aureobasidium pullulans* **(de By.) Arn.**
Pythium **sp.**
Rhizoctonia solani **Kühn. =** *Corticium solani* **(Prill. and Del.) B. and G.**
Rhizopus nigricans, **(Fr) Ehr. + various** *Mucor* **including** *Mucor hiemalis* **Wehmer**
Stemphylium botryosum **Wallr. =** *Pleospora herbarum* **(Fr.) Rabh.**
Stemphylium vesicarium **(Wallr.) Simmons =** *Pleospora* **sp.**

These mainly affect the fruit of field tomato crops, particularly those that are not staked. Their frequencies vary. *Alternaria* spp. are always present in large numbers, together with *Mucorales* to a lesser degree. The severity of attacks of other micro-organisms appears linked to a particular cultural system. This is the case for *Geotrichum candidum, Fusarium* sp., *Colletotrichum coccodes*. Others are always present in low proportions (*Stemphylium botrysum, S. vesicarium* etc.).

It should be noted that, in many cases, these occur alone or in combination with other fungi or bacteria (*Erwinia* sp., *Bacillus* sp.).

SYMPTOMS

See **241, 247, 248, 249, 250, 251, 252, 253, 258, 259, 260, 263, 267.**

PRINCIPAL CHARACTERISTICS

Sources: all these micro-organisms are saprophytic fungi capable or survival in the soil, on plant debris of the tomato or other plants.

Dissemination: by wind, rain (splashes and particles of soil), on contact with the soil or on contact with fruit already affected.

Penetration: for most of these organisms, via numerous wounds (growth splits, sunburn, blossom end rot, insect bites, various damage, etc.) and for some, directly through the cuticle of the fruit (*Alternaria* spp., *Colletotrichum coccodes*).

Conditions encouraging development: the agricultural and climatic conditions have a strong effect on the type and stage of appearance of the diseases. The damage is particularly severe following heavy rain. Free water present on the fruit or between the fruits for several hours encourages development. Also, water after a period of drought may cause numerous growth splits through which the pathogens can enter the fruits. Poor use of irrigation can cause the same situation.

These fungi rarely attack green fruits (in general, when they do, they remain 'latent' until the fruits start to ripen) but affect mainly ripe fruits, and the attack is especially severe on over-ripe fruit.

CONTROL METHODS

During cultivation: avoid excessive over-head irrigation, particularly during the evening.

Carefully dust the fruit with the following fungicides*: iprodione, chlorothalonil, imazalil.

Avoid harvesting fruit when over-ripe, as the risk of damage is high.

Next crop: a recommended treatment strategy adopted for tomatoes for processing is as follows:

For a single harvest

- Choose resistant varieties: Cannery Row, Earlymech, Lima, etc.
- Follow agricultural techniques carefully: divide beds into sections and water when required using a tensiometer to determine water requirement.
- On flowering: use an approved fungicide (once)
- At the first ripe fruit and every 15–20 days (three treatments): use an approved fungicide.

For multiple harvesting

- Keep to the number of harvests for the variety selected: Petogro, Rio Grande, Coudoulet, Europeel, Fl Lerica, Macero 2, etc.
- Cultivate the crop as for single harvested crops.
- Apply an approved fungicide treatment on flowering.
- A staggered harvest requires more treatments (four to five).

In certain situations, where possible, the use of plastic straw (to isolate fruit from the ground) or metal grids in the shape of an inverted U (supporting the plants) will reduce damage.

IMPORTANT NOTE

* Approval of fungicides may vary in different countries and local labels should be consulted before use.

Fungi (and nematodes) attacking the roots

DESCRIPTION 13

Pyrenochaeta lycopersici Gerlach

Causes corky root disease or brown and corky root rot

SYMPTOMS

See **140, 150, 151, 153.**

PRINCIPAL CHARACTERISTICS

Sources: this pathogen can penetrate to the deepest layers of soil, and can colonize root systems of cultivated plants (alternative hosts) such as melon, lettuce, aubergine, beans, cucumber and those of several weeds. It survives for a long time (several years) in the soil in the form of chlamydospores, on brown roots which are not corky. Mono-culturing of tomatoes increases the concentration of inoculum of the soil.

Strains adapted to cold, temperate and warm conditions can survive in the same soil and attack at different times of the year.

Very slow growth in the soil.

Dissemination: possible via all substrates, agricultural implements.

Optimum temperatures for development in the soil: 15–20°C for the northern European strains. There are in fact several strains with different temperature optima depending on their origin. Strains originating from the Mediterranean basin (Tunisia, Libya) are still pathogenic at temperatures of 26–30°C.

CONTROL METHODS

During cultivation: there are no effective ways of eliminating the pathogen present on the roots of the plants without causing definite damage to these. In an effort to keep plants alive as long as possible:

- Earth them up to encourage the growth of new roots which could supplement the older, necrotic roots. In soil-less culture (on peat or pozzolane + peat), with a serious attack, peat can be applied locally to the collar in order to encourage additional rooting.
- Spray with water at the hottest times of the day to avoid excessive evaporation that is not compensated for by up-take, and which can lead to the desiccation and death of the plants.
- Watch irrigation carefully: if the plants wilt, root damage by *Pyrenochaeta lycopersici* could be to blame, or it could be lack of water. Too often, growers tend to increase irrigation which leads to further root death.
- At the end of cultivation, carefully lift plants to remove as many damaged roots as possible.

Next crop: plants can be grafted on to root stocks (KNVF*, Hires) or resistant hybrids used (if the agronomic quality of these is suitable, see Appendix 2) if you prefer not to disinfect the soil. In the case of field crops, rotation will not help as the fungus remains in the soil for a long time despite the growth of other crops.

Tests have shown that by increasing the volume of the mounds, the damage to the roots is reduced.

IMPORTANT NOTE

* KNVF: root stock resistant to pathogens responsible for corky root disease, root galls due to nematodes, Verticillium, Fusarium (see list in Appendix 2).

Disinfection of soil

You can use steam (few sites are equipped), methyl bromide and/or chloropicrin (in countries where this is permitted). These are the most effective products. Dazomet or metham sodium also give perfectly acceptable levels of control.

Solar disinfection of the ground (solarization) could be considered; spectacular results have been recorded in some Mediterranean countries. This technique consists of covering the soil (very well prepared and moist) with a 25–40 μm thick polyethylene film, and keeping it in place for at least 1 month at a very sunny time of the year.

Disinfection of substrates of soil-less crops

At present little information on the disinfection of substrates is available. Steam treatment is effective if a suitable system for treatment can be used.

Metham sodium cannot be used for organic-based substrates, but could be used for mineral substrates (sand, perlite, pozzolane). Work on a dry substrate, repeating the treatment several times, then leach well.

Methyl bromide can be used; method of use to be established with the user.

Formaldehyde has been successfully used for disinfecting pozzolane and peat (by impregnation of the substrate with 3% commercial formaldehyde).

To propagate plants, use healthy (disinfected) substrates, and avoid placing roots next to the nursery soil, especially if this has not been disinfected (contamination may result from contact).

Colletotrichum coccodes (**Wallr.**) **Hugues**
Colletotrichum atramentarium (**Berk. Br.**) **Taub.**

Causes root necrosis, anthracnosis and black dot

SYMPTOMS

See **152, 251, 252, 253**.

PRINCIPAL CHARACTERISTICS

Sources: on roots in the form of microsclerotia (acervuli of the pathogen also form on the roots). The disease can develop on roots of other secondary hosts such as aubergine.

Dissemination: during working of the soil by agricultural implements. The spores can be transferred by water over considerable distances.

Conditions for development: optimum temperatures 22–24°C. The fungus prefers light soil, as heavy soil tends to retain water and hence discourage the preservation of the microsclerotia.

Not all pathologists regard this fungus as a parasite. In France, it has been known to occur in the following conditions:

- In the presence of *Pyrenochaeta lycopersici* on the same root systems, each apparently developing on different parts of the root.
- Alone on root systems which show only characteristic changes, without the presence of corky sleeves.
- On root systems which also have *Rhizoctonia solani*.

It is relatively rare in soil-less crops, although it has been reported several times in NFT crops. This parasite is fairly common in open field crops in the south-east of France.

CONTROL METHODS

When attacks are serious, treat as for *Pyrenochaeta lycopersici* (see description 13).

DESCRIPTION 15

Spongospora subterranea (**Wallr. J. Johnst**)

Causes root galls

SYMPTOMS

See **143, 156, 157, 158, 159.**

PRINCIPAL CHARACTERISTICS

This fungus is rarely seen in crops, although it sometimes causes spectacular symptoms which have no apparent effect on the plants, whose vigour and root production is scarcely affected. This fungus does not seem able to survive for a long time in soil except on tomato roots.

CONTROL METHODS

There is no real method of treatment; at the end of cultivation, remove as many affected roots as possible.

DESCRIPTION 16

Fursarium oxysporum Schlect f.sp. *radicis-lycopersici* (Jarvis and Shoemaker) (F.O.R.L.)

Causes root rot or crown and root rot

SYMPTOMS

See **123, 128, 144, 145, 147, 148, 149, 168, 169, 170, 234.**

PRINCIPAL CHARACTERISTICS

Sources: in many substrates and in soil, on plant debris or in the form of chlamydo-spores (resistant spores). Aubergine and pepper may also be affected by this pathogen, which causes root necrosis in its hosts.

Dissemination: very easy via the conidia (numerous spores on cankers located at the collar of the plants on rock wool blocks) or particles of contaminated soil. Both can be carried by water (irrigation water, splashes), wind and air currents. This parasitic fungus was introduced into several crops through contaminated compost or plants.

Conditions encouraging development: the parasite appears able to colonize plants irrespective of soil temperature. Previously, it was considered as a parasite of relatively cold soils or substrates, mainly affecting plants of early crops and the colder parts of greenhouses (optimum temperatures around 18–20°C). In some countries, this disease has occurred several times in summer when the temperatures were above 26°C. The severity of the disease depends also on the aggressiveness of the strain and appears worse when the roots have been subjected to excess water or low temperatures at any time in the culture.

This *Fusarium* is capable of rapid colonization or recolonization of recently disinfected soils or substrates.

CONTROL METHODS

During cultivation: at present, there is no effective treatment for this disease. To help plants survive as long as possible, the following measures should be carried out:

- Earth up plants to encourage the production of new roots which could supplement the older, necrotic roots. In soil-less crops (on peat or pozzolane + peat), peat can be applied to the collar to encourage further rooting.
- Treatments of benomyl-based fungicide may be applied locally by drenching the base of the plants (this method of application gives a better distribution of the product by the root system) or via the nutrient irrigation system. Their effectiveness is not consistent and strains of *Fusarium oxysporum* f.sp. *radicis-lycopersici* resistant to benomyl have already been reported following repeated use. This product can also carry some risk of phytotoxicity, in particular on rock wool.
- Systematically remove dead plants, in particular their root system and basal stem portion (the fungus sporulates well on cankers which are often present on the collar).

Next crop: avoid rotating with aubergine or peppers; use lettuce instead which is not sensitive.

The entire operation must be disinfected to eliminate the 'propagules' which can contaminate plants of the next crop.

The soil may be 'sterilized' either by steam or using a fumigant such as methyl bromide; the duration of effectiveness of this treatment is fairly short as the fungus recolonizes disinfected soil very quickly. To remedy this, and for steam disinfection, some authors recommend Captafol at the dose of 26.8 kg active matter/hectare on treated soil.

The surfaces of the internal structures of the greenhouses should also be disinfected. Formaldehyde can be used, which is an excellent surface disinfectant with rapid action when in vapour form. A 2% formaldehyde solution (of commercial formaldehyde at 38%) may be sprayed at the rate of 200–250 l/hectare before and after soil disinfection. After treatment, the greenhouse should be closed for at least 48 hours (see also treatment methods for *Fulvia fulva*, description 8).

To produce plants, use healthy substrates and avoid placing roots next to the nursery soil. Do not use plants from contaminated sources.

Resistant root stocks are now being developed and will probably be available soon, as will resistant varieties.

Disinfection of substrates of soil-less crops is not commonly practised; it is better to discard used substrates. For this type of crop, cover the soil with a plastic film to isolate the crop and avoid possible contamination. For the same reason, tanks of nutrient solution and water storage reservoirs should be covered.

Other fungi responsible for root symptoms

Poor cultural conditions, in particular excess water or salinity, and soil temperatures which are too low, can lead to changes in the root system which often result in colonization by secondary pathogens or invading fungi such as *Colletotrichum coccodes*, *Fusarium* sp., *Olpidium* sp., *Rhizoctonia solani*, various Pythiaceae (*Pythium ultimum*, *Pythium aphanidermatum*, *Phytophthora nicotianae* var. *parasitica*, etc.).

Of these fungi, several Pythiaceae and *Rhizoctonia solani* are also responsible for damage (sometimes severe) during propagation or after planting. Symptoms include a lack of growth, damping off, and necrosis of the roots and/or collar. Also, several Pythiaceae, in particular *Pythium*, are found in substrates (permanently wet) of soil-less crops where the conditions are perfect for their development and where they cause harmful root loss.

Many fungi also attack the collar of the plants; further information is given in the section on these.

Meloidogyne spp. (gall nematodes)

Causes root gall

SYMPTOMS

See **130, 138, 154, 155**.

Depending on the regions and countries, various species of *Meloidogyne* can occur:

- *Meloidogyne arenaria*
- *M. javanica*
- *M. incognita*
- *M. hapla*

PRINCIPAL CHARACTERISTICS

Sources: egg masses protected by a viscous coating which can survive for more than 2 years. These nematodes are highly polyphagous and attack numerous vegetable and fruit crops (artichoke, aubergine, pepper, cucumber, melon, lettuce, almond trees, peach trees, olive trees, etc.).

Dissemination: possible via contaminated plants, tools and irrigation water.

Conditions encouraging development: they like relatively high temperatures (18–27°C) which are found in light sandy soils, except for *M. hapla* which is adapted to lower temperatures.

CONTROL METHODS

During cultivation: there is no effective treatment. However, plants should be earthed up and sprayed with water at the hottest times of the day.

Remove and destroy the root system of affected plants because these contain a high number of nematodes.

Next crop: dig deep in mid-summer, when nematodes exposed to the sun will dry out and die.

Use healthy plants.

Resistant plants are available such as KNVF root stock (see Appendix 2): these are highly resistant to one or more species of *Meloidogyne*. No variety is resistant to *Meloidogyne hapla*. Avoid growing a resistant variety too often on the same ground, as the nematodes can adapt to this variety and attack it.

Soil disinfection is often undertaken. Use either steam, methyl bromide alone or in combination with chloropicrin or dichloropropane-dichloropropene (DD). Other fumigants are sometimes applied (dazomet, metham sodium).

Biological treatments have been applied in recent years, using *Arthrobotrys* fungi which are predatory to the nematodes (in particular *Arthrobotrys irregularis*). This must be applied in lightly contaminated soil, very early (at least two months before the crop). The fungus is spread over the ground and is slightly buried. The fungus can be used only if the following conditions are fulfilled:

- The pH of the soil is above 6.4, soluble salts less than 2%, organic matter more than 8%.
- Healthy plants.
- Fungicide treatments must not be applied to the soil.

In the case of heavily contaminated land, disinfect before establishing the fungus.

Fungi attacking the collar and sometimes the stem

DESCRIPTION 19

Didymella lycopersici Kleb
Causes didymella cankers

SYMPTOMS

See **167, 188, 192**.

PRINCIPAL CHARACTERISTICS

Sources: in the soil, on plant debris, on stakes and supports which have been in contact with diseased plants.

Dissemination: (considerable, from sporulating cankers on the collar and stem): by tools during cutting or soil working, by air currents and splashes. Mainly via the conidia (from pycnidia, asexual reproductive organs) rather than via ascospores (from sexual reproduction). The fungus can be transmitted by seeds, but this is rare. The fungus would not survive more than nine months on seeds.

Conditions encouraging development: this fungus is capable of attacking at temperatures which vary from 15–28°C with an optimum of 19-20°C. It prefers the high humidities of protected crops. It appears to attack most severely in disinfected soils which have been recontaminated. It establishes itself quickly through wounds, in particular bud and leaf removal wounds.

CONTROL METHODS

During cultivation: in the case of a slight attack, with only a few diseased plants, remove these from the greenhouse and destroy immediately; avoid overhead irrigation, air and ventilate protected crops as much as possible.

For severe attacks, follow the above measures and carry out treatment either locally to the collar of the plants (using benomyl or carbendazim) or stem cankers (by painting with benomyl, iprodione, vinclozolin, with or without paraffin oil), or generally using these products or with maneb, Mancozeb or fenarimol.

Destroy plant debris, and do not dig this into the ground. Remove any items which have been in contact with the diseased plants and disinfect (particularly stakes and tools) using bleach/disinfectant or a formaldehyde solution (see treatment methods for *C. michiganensis*, description 3).

Next crop: use a pathogen-free substrate for plant raising, as *Didymella* sometimes attacks young plants.

If the attack has been severe, disinfect the soil by steam or with a fumigant (see treatment methods for *Pyrenochaeta lycopersici*, description 13).

KNVF root stock is resistant to *Didymella* and can be used.

DESCRIPTION 20

Phytophthora nicotianae var *parasitica* Dast

Causes root loss and collar canker

SYMPTOMS

See **129, 164, 165, 273, 281, 283**.

PRINCIPAL CHARACTERISTICS

Sources: persistent in the soil. This fungus has many hosts although it shows different degrees of aggression towards different ones (various vegetable crops and flowers).

Transmission: via the substrate, water after splashing from the soil, or via contaminated water reservoirs and irrigation channels.

Conditions encouraging development: it mainly attacks plants which have been subjected to stress as their roots (excess water, or soil or substrate temperatures too low), in particular in the weeks following planting (always a critical period for young plants). It appears to develop at temperatures between 15°C and 26°C.

CONTROL METHODS

The methods described here also apply to the pythiaceae attacking the tomato.

During cultivation: during propagation, heat and restrict irrigation if irregularities have been found (shortages or excess).

Also, immediately apply anti-pythiaceae products to the root system and collar of the plants. Several methods are possible, depending on the situation:

- During propagation, immediately drench the entire substrate with a fungicide solution.
- For soil crops, without individual plant irrigation, the fungicide solution can be applied (by direct jet) to the base of diseased plants and healthy neighbouring plants.
- For soil or soil-less crops, where water and/or nutrient solutions are applied locally to the base of the plants (drip system), the fungicide can be applied via the irrigation system. In certain substrates, this method of application of fungicide does not guarantee a good distribution through the entire root system of the plants.

Several fungicides* are used against pythiaceae:

- Metalaxyl.
- Furalaxyl.
- Propamocarb HCl.
- Etridiazole.

It is difficult to recommend a dose for each product, as this will differ depending on the method of cultivation of the crop. For soil-less crops, in particular in NFT, and for certain substrates such as rock wool, the doses must be reduced as there are considerable risks of phytotoxicity.

Remove dead plants quickly.

Attacks on fruit can be treated by spraying the appropriate fungicides; this treatment is not always successful on unstaked plants as often the under-side of the fruit is not treated. Avoid keeping the soil moist for too long, and stop fruits dipping into the water.

Next crop
During propagation:

Use a healthy disinfected substrate. Do not mix with sand or soil collected from the field, as this can be contaminated.

Avoid placing the roots of young plants next to the ground unless this has been disinfected; arrange the plants on benches or a plastic film which can be covered with a layer of peat. Replace this often.

IMPORTANT NOTE

* Approval of fungicides may vary in different countries and local labels should be consulted before use.

The medium used for sowing can be sprayed with a preventive treatment of fungicide solution (1 ml liquid Cryptonol + 1 ml Previcur N per litre of water).

On planting:

Avoid planting in soil which is too cold, and avoid adding too much water to the neck of the plants.

In soil-less crops, contaminated substrates must be replaced or disinfected.

Do not use irrigation water, especially if taken from canals or 'pools' which may have been contaminated.

DESCRIPTION 21

Rhizoctonia solani Kühn

Causes root and neck rot

SYMPTOMS

See **166, 274, 275.**

PRINCIPAL CHARACTERISTICS

Sources: in the soil in the form of mycelium or sclerotia. This is a highly polyphagous pathogen (over 25 hosts are known), capable of attacking and surviving on debris from various plants (lettuce, cucurbitaceae, aubergine, leeks, beans, several weeds, etc.).

Contamination: possible via horticultural substrates or potato tubers remaining from the previous crop.

Conditions encouraging development: the fungus is present in particular in market-garden soils which have grown vegetable crops for several seasons. It can develop in humid, heavy soil and in lighter, drier soil, at temperatures between 15°C and 26°C.

CONTROL METHODS

During cultivation: *Rhizoctonia* rarely requires action during cultivation except in the case of attacks during propagation or after planting. Apply quintozene (PCNB) or thiophanate-methyl locally to the base of the plants. The effectiveness of this treatment is not consistent.

Eliminate diseased plants and plant debris during and at the end of cultivation.

Avoid excess irrigation, particularly in heavy soil.

Damage to fruit does not usually require treatment.

Next crop: use healthy substrates and avoid placing young plants on the soil during propagation. In the case of heavily contaminated soil, either disinfect before planting with a fumigant such as dazomet or metham sodium, or spread quintozene on the ground.

In countries where it is permitted, solarization (see treatment methods for *Pyrenochaeta lycopersici*, description 13) is an economical and effective method of eliminating this fungus which is often present in the surface layers of the soil.

DESCRIPTION 22

Sclerotinia sclerotiorum (Lib) de By

Causes cankers to the collar and stem

SYMPTOMS

See **174, 175, 189, 191.**

PRINCIPAL CHARACTERISTICS

Sources: survives for several years, through numerous large sclerotia or mycelium present in plant debris abandoned on the soil. This is a highly polyphagic fungus which can be found on over 360 different hosts, particularly on lettuce, aubergine and pepper, which are often rotated with the tomato.

Contamination occurs in two ways:

- Through mycelium from sclerotia, at soil level.
- Through ascospores from apothecia (organs for sexual reproduction of the fungus formed from sclerotia) which contaminate the aerial part of the tomato and disseminate the disease over several hundred metres.

Conditions encouraging development: this fungus is encouraged by relatively low temperatures between 15°C and 18°C (minimum 5°C, maximum 30°C) and high relative humidities either at soil level or within the crop canopy. It is very sensitive to carbonic gas, which is why it is localized in the first few centimetres of soil. Light soil, rich in humus, is most suitable for its development.

CONTROL METHODS

During cultivation: remove dead plants carrying sclerotes.

Reduce humidity in greenhouses by ventilating as much as possible.

Avoid overhead irrigation.

Apply a solution containing one of the following fungicides* to the neck of the plants: benomyl, thiophanate methyl, iprodione, vinclozoline, procimidone.

Cankers on the stem can be painted in the same way as those caused by *Botrytis cinerea* (see description 7).

If air-borne contamination appears to be occurring, spray with iprodione + neutral oil.

At the end of cultivation, remove and destroy the affected plants and their sclerotia.

Next crop: many vegetable crops may be attacked (lettuce, aubergine, beans, cucumber, melon, etc.) and these are often rotated with the tomato, so if the land is heavily contaminated, disinfect the soil. Many fumigants (methyl bromide, dazomet, metlam sodium, etc.) and steam can be used.

Use healthy plants.

IMPORTANT NOTE

* Approval of fungicides may vary in different countries and local labels should be consulted before use.

Vascular fungi

DESCRIPTION 23

Fusarium oxysporum f. sp. *lycopersici* (Sacc.) Sn. and H.

Two strains in France: 0 (previously 1); 1 (previously 2)
Causes Fusarium wilt

SYMPTOMS

See **125, 210, 218, 219, 220, 221, 222, 224.**

PRINCIPAL CHARACTERISTICS

Sources: survives for very long periods in the soil and on plant debris because of very resistant spores, the chlamydospores. The fungus colonizes the soil at a great depth, below 80 cm.

Transmission: possible via compost, soil particles (sometimes over long distances, especially in soil-less crops), water, agricultural tools and implements, crop debris and insects such as millipedes.

Conditions encouraging development: this fungus prefers to colonize during the warm parts of the year. The optimum temperature for development is 28°C. It prefers sandy and acidic soils. Plants are particularly sensitive to the disease when suffering from a deficiency of nitrogen, phosphorus and calcium, and when subjected to short days and low light levels. Varieties resistant to Fusarium wilt sometimes show signs of this disease, particularly when severely attacked by nematodes (*Meloidogyne* spp.) or when plants have suffered root suffocation (in these situations, the strain concerned may not have adapted).

CONTROL METHODS

During cultivation: there is no effective treatment for this disease. Benomyl or thiophanate methyl are sometimes applied to the base of plants; the cost of successive treatments (often ineffective) is very high in relation to the results obtained. Remove diseased plants and crop residue where possible during and after cultivation.

Next crop: during propagation, use healthy substrates.

The genetic solution is by far the most satisfactory method of controlling this disease. There are many varieties resistant to strain 0 and an increasing number resistant to both strains (see Appendix 2). Grafting on to the same stocks as those used to combat *Pyrenochaeta lycopersici* can be considered as these are resistant to this *Fusarium* (see description 13).

When a sensitive variety is preferred, or if there is no resistant variety to meet the requirements of the grower, the soil may be disinfected instead of grafting undertaken. Disinfection is not always completely successful: it depends on the fumigant applied and above all on the precautions taken after disinfection to avoid early recontamination by the pathogen. Even if great care is taken, disinfection will last for only one season.

The best fumigants for disinfecting the soil are chloropicrin, methyl bromide and a combination of the two. For protected crops, some authors recommend covering the ground completely to avoid recontamination (this measure is essential for soil-less crops). In addition to the soil, the structures, equipment, nutrient solution tanks (which must be covered) and nutrient solution feed circuit must be disinfected. To do this, use a 3% formaldehyde solution or bleach/disinfectant, which can also be used to disinfect the substrates of soil-less crops.

Avoid excessive phosphorus and magnesium applications, and add nitrogen as nitrate rather than ammonia.

Liming will attenuate the effects of Fusarium wilt.

DESCRIPTION 24

Verticillium dahliae Kleb. *(Verticillium albo-atrum)* Reinke and Berth. (more Nordic variety)

Causes Verticillium wilt

SYMPTOMS

See **127, 225, 226, 227, 228, 229.**

PRINCIPAL CHARACTERISTICS

Sources: this fungus can survive in the soil for a very long time because of resistant structures (microsclerotia) and via numerous host plants, cultivated or weeds (aubergine, black nightshade, amaranth, etc.). Repeated tomato crops attract strains which are particularly aggressive to this host.

Transmission: possible via compost, tools and plant debris. The conidia are easily disseminated in greenhouses by air currents and by water splashes.

Conditions encouraging development: for maximum aggressiveness, this fungus needs relatively cool temperatures (20–23°C). In France, it mainly attacks in the spring and autumn. Short day length and low light will make plants more sensitive to the disease; similarly, early planted crops with reduced foliage and trusses setting are particularly sensitive.

CONTROL METHODS

During cultivation: there are practically no methods of treating this disease during cultivation; however, in the case of an early soil-grown crop, applications of benomyl or carbendazim (at the rate of 0.5–1 g active ingredient per plant) to the base of each plant are recommended as soon as symptoms appear. These treatments (to be repeated every 3 weeks) will limit the development of the fungus in the plants until temperatures reach levels (above 25°C) which inhibit its development. In the case of soil-less crops, doses must be reduced as there are risks of phytotoxicity (yellowing and necrosis of the edge of the lamina of the leaflets).

Next crop: the best solution is to use a resistant variety (see Appendix 2 for list of varieties with a high level of monogenic resistance).

Other treatment methods (grafting, disinfection of the soil and substrates) used to combat Fusarium wilt can be used against Verticillium wilt; these methods often have the same limitations when used against the latter (see description 23).

8. Viruses

DESCRIPTION 25

Tobacco mosaic virus (**TMV**)

(Four strains known: 0, 1, 2, 1–2)
Causes tomato mosaic

SYMPTOMS

See **15, 16, 23, 33, 39, 40, 42, 44, 82, 239, 278, 279, 287, 288.**

PRINCIPLE CHARACTERISTICS

Transmission: very easy, by contact during pricking out, pruning and harvesting fruits (via tools, clothing), via the seeds, water (through the roots), especially in soil-less crops.

Source: in seeds, plant debris, the soil, compost, etc.

Deactivation: by heating the seeds to 80°C for 24 hours (heat therapy).

CONTROL METHODS

During cultivation: during cultural work, disinfect the hands and tools after working in an infected plot.

Remove diseased plants and tomato debris buried in the soil as this disease can be transmitted by root contact.

Next crop: disinfecting the soil with steam (100°C for 10 minutes) is effective in deactivating the virus, as is the use of methyl bromide (75 g/m^2).

Use healthy seeds.

Resistant hybrids which are effective against the four known strains (see Appendix 2) are available in several varieties, in particular for the greenhouse.

Use of direct sowing (for open field crops) avoids contamination caused by handling plants grown in the nursery.

Inoculation of the plants with a weak strain has been an effective method of protecting susceptible varieties: this method has recently been abandoned as there are now many resistant varieties.

Potato virus (**PVX**)

This virus is also transmitted by contact. In most cases it causes necrotic spots on the leaves of the tomato. When combined with tobacco mosaic virus, necrosis can appear on the petioles of varieties which are susceptible to TMV.

DESCRIPTION 26

Cucumber mosaic virus (CMV)

'Common' strains: mottling, fern leaf

Strains with necrotic ARN satellite: necrosis

Causes fern leaf, mottling and necrosis of the tomato

SYMPTOMS

See 5, 12, 17, 24, 37, 41, 43, 135, 202, 272.

PRINCIPAL CHARACTERISTICS

Transmission: by aphids in a non-persistent way. By feeding on an infected plant, the aphid immediately becomes a carrier of the virus, i.e. it can transmit the virus and thus the disease to one or more other plants immediately after feeding and for some hours. Many types and species of aphid are vectors, including *Myzus persicae* and *Aphis gossypii*.

The disease can be present in a crop without any aphids being observed on the plants.

Source: attacks many cultivated and wild plants: it is the latter which ensure the survival of this virus through the winter.

CONTROL METHODS

During cultivation: there is no cure for this virus. If only a few plants are affected, remove them.

Insecticide treatments will have no effect, or only a slight effect, on the spread of this virus which can be transmitted in a very short feeding time by aphids coming from outside the plot, even before the aphicide has had time to act.

Next crop: there is no really effective preventive measure. However, in order to limit or delay contamination by aphids, the following may help:

● Protect young plants by 'food protection' type grilles or with non-woven fabric type Agryl P17. This advice applies in particular to very early tomato crops which are planted in the autumn at a time when aphids are still very numerous.

● Weed the plots and their edges.

● Cover crops with a plastic film (heat transparent or opaque) which eliminates aphid populations (this measure applies only to open field crops).

DESCRIPTION 27

Potato virus Y (PVY)

Causes mottling and necrotic spots on tomatoes

Two types of strain: mottling strains; mottling and necrotic spots (the most harmful)

SYMPTOMS

See 38, 45, 46, 47, 80, 81, 82.

PRINCIPAL CHARACTERISTICS

Transmission: by aphids in a non-persistent way. Several species are vectors, including *Myzus persicae, Aphis citricola* and *Aphis gossypii.*

Source: the virus attacks cultivated (potato, pimento) and wild Solanaceae (black night-shade = *Solanum nigrum* and woody nightshade = *Solanum dulcamera*) and other weeds such as purslane (*Portulaca oleracea*) and groundsel (*Senecio vulgaris*).

This virus also affects the potato and the pimento; strains attacking these hosts do not appear to attack the tomato as severely.

CONTROL METHODS

The treatment methods are those used to treat viruses transmitted by aphids in the non-persistent manner (see cucumber mosaic virus, description 26). Particular attention should be paid to protecting nurseries, especially in the case of very early sowings in September; at that time, aphids carrying the virus are still present and the risk of contamination is high.

DESCRIPTION 28

Alfalfa mosaic virus (AMV)

Causes necrotic mottling of the tomato

SYMPTOMS

See **48, 201, 271, 277.**

PRINCIPAL CHARACTERISTICS

Transmission: by aphids in the non-persistent manner, as for cucumber mosaic virus; many types and species of aphid are vectors, including *Myzus persicae* and *Aphis gossypii*.

CONTROL METHODS

This virus is encountered in several plants and is rarely serious except in the case of a tomato field near to an alfalfa field—avoid placing crops near to alfalfa.

See the treatment methods used to combat cucumber mosaic virus (description 27).

DESCRIPTION 29

Tomato yellow leaf curl virus (TYLCV)

Causes yellow leaf curl disease

SYMPTOMS

See **11, 14, 22**.

PRINCIPAL CHARACTERISTICS

Transmission: the tobacco white fly *(Bemisia tabaci)* is the only known carrier. *Trialeurodes vaporariorum* (the white fly which transmits lettuce and cucumber yellows) does not appear to transmit this disease. Transmission by seeds or contact is not possible.

Source: several weeds will accommodate and conserve the virus during the warm periods of the year, in particular *Malva nicaensis* and *Datura stramonium*. Tobacco is the only other cultivated plant to be affected by the virus.

CONTROL METHODS

During cultivation: there is no really effective treatment. Repeated applications of anti-whitefly treatments* (bioremethrine, deltamethrine, cypermethrine, pyrimio-phos-methyl, dichlorvos, methomyl, etc.) have only a slight effect in limiting the spread of the disease.

IMPORTANT NOTE

* Approval of insecticides may vary in different countries and local labels should be consulted before use.

Next crop: as before, there is no really effective treatment. To give an acceptable level of protection, the following measures can be combined where possible:

- Protect the young plants in the nursery from whitefly by sustained chemical treatment (two to three applications per week) or by covering them with a non-woven fabric type Agryl P17. The latter, depending on location, can bring some problems (lack of light, excess humidity, high cost).

- Cover the ground with fresh straw or a film of yellow polythene. This method appears to delay infection by 2–4 weeks.

- Remove plants containing the virus (weeds), in particular near tomato crops. Avoid planting tomatoes near tobacco crops.

- Plant tomatoes out of season (at times of the year when the vectors are still rare) or intersperse these with plants attractive to the vector (but not sensitive to the virus) in order to reduce the frequency of the disease.

9. Mycoplasma from the 'aster yellows' group

DESCRIPTION 30

Causes stolbur

SYMPTOMS

See 7, 9, 28, 29, 30, 31, 35, 53, 54.

PRINCIPAL CHARACTERISTICS

Transmission: via Cicadellidae (leaf hoppers). Several vector species have been reported, in particular *Hyalestes obsoletus* which feeds on bindweed (*Convulvulus arvensis*). The disease can also be transmitted by grafting. It is not transmitted by contact or by seeds.

Source: leaf hoppers can spend the winter as eggs or adults on intermediate hosts, often perennial herbaceous or woody plants, for example the bindweed in the case of *Hyalestes obsoletus*.

Conditions encouraging development: these are not known. The disease attacks sporadically and no explanation has been given for this feature.

CONTROL METHODS

During cultivation: there is no treatment for this disease. Once the symptoms have appeared, it is too late to act.

Next crop: there is no treatment to control the disease completely. The measures suggested here will only delay and limit contamination as far as possible (late attacks are less harmful):

- Remove weeds in the crop and in the vicinity.
- Avoid placing young plants near crops already affected; these will have been produced under protection either in greenhouses or protected by a non-woven polypropylene cloth (type Agryl P17).
- Apply insecticides early to tomatoes and woody or herbaceous plants at the edges of the plot. Most insecticides used in vegetable crops will be effective against leaf hoppers.

Measures to be taken to avoid some non-parasitic diseases

DESCRIPTION 31

The measures to be taken to limit the development of non-parasitic diseases are often related to their causes (see Part one of this book). In many cases, poor climate control and/or unsuitable agricultural conditions, etc. are to blame. This is the case, for example, for frost damage, leaf curl, root suffocation and various deficiencies. To treat such damage, the basic faults must be corrected and/or conditions of stress minimized.

Some diseases, with more complex causes, require more specific measures, which are described below:

Silvering
(See page 36 and **34, 49, 50, 51, 52**).

Treatment
None during cultivation; however you can leave axillary buds on plants which may not be showing the symptoms. In this case, remove the affected apex.

- Avoid growing varieties which have been shown to be particularly sensitive to silvering in cultural tests.

Yellow collar
(See page 141 and **286**)

Treatment
- Avoid high temperatures in the greenhouse and after harvesting.
- Avoid severe pruning.
- Balance the manuring.
- Harvest when colour is turning and keep the crop at temperatures below 22°C.

Oedema
(See pages 25–53 and **84, 86, 87**).

Treatment
- Ventilate (watch tightly sealed greenhouses which may require additional ventilation before nightfall).
- Heat if the increase in humidity is caused by a fall in air temperature.
- Watch for the development of *Botrytis cinerea* on damaged tissue.

Blotchy ripening
(See page 141 and **287, 288**).

Treatment
- Choose varities without this problem.
- Keep night-time temperatures lower than day-time temperatures.
- Avoid excessively high temperatures during the day.
- Feed the plants properly, in particular with potassium during the shorter days.

Sterile mutants
(See page 121 and **10, 18, 22**).

Treatment
No treatment during cultivation. Eliminate the mutants on planting or afterwards.

Phytotoxicity
(See pages 23, 39, 49, 54, 71, 96 and **6, 8, 19, 20, 21, 61, 62, 63, 64, 65, 66, 67, 88, 89, 90, 131, 132, 133, 134, 180, 204, 289, 290, 291, 292**).

Treatment
Determine the cause of the phytotoxicity.

- Prevent its recurrence.
- Do not remove plants immediately. Grow normally and watch their development. This will depend above all on the dose and residue of the products concerned.
- No specific measures can be recommended.

304

305

306

307

Appendix 1

Damage by principal pests of the tomato

SPIDER MITES

Tetranychus urticae
Tetranychus cinnabarinus (less common)
Aculops lycopersici (Vasates lycopersici)

	Tetranychus urticae	*Aculops lycopersici*
Appearance of pests	Globular spider mite yellowish-green, 0.3–0.5 mm long. Spherical translucent eggs, 0.1 mm diameter	Minute straw-yellow spider mite (invisible to the naked eye), 0.12–0.15 mm long
Damage	Growth stops. Small yellow punctures on leaflets (304), presence of numerous silky webs (305, 306)	Gleaming spots on stems. Bronze–green colour leaflets, drying and dropping of leaflets and leaves. Corky, cracked fruit (307)

Tetranychus cinnabarinus
(same as *T. urticae*, colour crimson)

LEAF MINERS
Liriomyza bryonia
Liriomyza strigata
Liriomyza trifoli (very aggressive)

Appearance of pest
Adult: highly mobile 'fly', 2 mm long, yellow and black in colour (**309**). Larva yellow, 1 mm long, moving in the thickness of the leaf (**310**); pupa drum-shaped.

Damage
Minute yellowish punctures (feeding punctures) and numerous curving galleries (**308**) on leaflets which then dry out (**311**).

308

309

310

311

CATERPILLARS

Heliothis armigera (most common species)

Mamestra oleracea (caterpillar feeding on the tomato) and several other species can also settle

Appearance of pest
Butterfly with a wing-span of 35–50 mm (male grey, female orange-brown). Caterpillars 35–45 mm long, variable in colour (**313, 315**).

Damage
Perforation of leaflets (**312**). Holes on fruits (**313, 314**) causing premature ripening, hollow in fruit filled with fly specks (**315**).

12

13

14

315

316

317

318

319

APHIDS AND WHITEFLIES

Aphis gossypii
Aulacorthum solani
Macrosiphum euphorbiae
Myzus persicae
Trialeurodes vaporariorum (whitefly)

	Aphids	Whiteflies
Appearance of pests	Insects of low mobility, 1.5–2.5 mm long, colour variable depending on species (green, yellow, black, pink, etc.) (**317**)	Small winged insects, white, 1.2–1.5 mm long (**318**). Flat larvae, oval in shape, with cilia.
Damage	Growth stopped. Leaflets and leaves deformed and shrivelled. Honeydew produced, covered with sooty mould (**316, 319**)	Honeydew produced, covered with fumagine (**316, 319**)

Sooty mould is the black mould which covers the leaflets and fruits of tomatoes colonized by certain insects such as aphids and whitefly. The mould consists of several species of fungi (*Alternaria, Cladosporium, Penicillium,* etc.) which, when developing on the honeydew (a sugary excretion which is the nutritional base), often colour it black.

BROOMRAPE AND DODDER

A species of broomrape (*Orobanche ramosa*) and several species of dodder have been reported on the tomato; the plants affected are less vigorous and can sometimes die if these parasitic plants are allowed to develop. Their characteristics are as follows:

Broomrape
Suckers attached to single or ramified roots and stems, leaves in the form of scales, spiky inflorescence (320, 321).

Dodder
Suckers attached to young stems, twining leafless stalks, small flowers in bunches (322, 323).

320

321

322

32

Appendix 2

The resistance to disease and pests of principal varieties of tomato grown in the Mediterranean basin

Genetic treatment, based on the use of resistant varieties, appears to be ideal insofar as it eliminates or reduces the need for fungicides during cultivation. This in turn reduces the chemical pollution of crops and the environment. The efforts made by tomato breeders over the past 20 years have led to the creation of varieties which have greater resistance to many diseases.

However, experience in recent years shows that there are limits to the effectiveness of genetic treatment.

In certain cases, the level of resistance is inadequate. For example, resistance to blight (caused by *Phytophthora infestans*) controlled by gene Ph-2 is very low when the plants are etiolated.

In several cases, the resistance has been overcome by new strains. Thus in many countries, in some areas or crops, strain 0 (previously called strain 1) of *Fusarium*, controlled by gene 1, is no longer effective as strain 1 (previously called strain 2) is present, and only varieties with gene 1–2 are resistant. The most typical example of rapid adaptation of the pathogenic agent to the resistance genes concerned is that of *Fulvia fulva* in greenhouses, where the three genes *Cf-2*, *Cf-4*, *Cf-5* and recently *Cf-9* have been overcome in succession.

To ensure a more durable barrier to the pathogenic agent, selection is tending to include in the same variety two resistance genes controlling different mechanisms of reaction to infection. This is relatively easy with the tomato in which most resistance genes are dominant, which allows their culmination in F1 hybrids. For example, in hybrids which carry genes *Tm*-1 and *Tm*-2^2 at the same time, the former controls tolerance of and the latter hypersensitivity to the tobacco mosaic virus. In selections currently in progress, *Fulvia fulva* resistance is provided by gene *Cf*-6 combined with newly discovered genes.

The question is often posed: what are the actual mechanisms of the resistance, and how does it work? Several types of reaction have been observed:

- **Partial resistance** (or reduced sensitivity): the development of the disease is slowed, the symptoms are reduced in relation to the reaction of sensitive plants. Examples: genes *Ph*-2 (resistance to *Phytophthora infestans*) and *Cf*-6 (resistance to *Cladosporium*).
- **Tolerance:** in the case of viral diseases, where the pathogenic agent finds it more difficult to infect, migrates more slowly in the plant, multiplies more slowly, and where the plant shows no symptoms. Example: gene *Tm*-1 cited above.
- **Immunity:** a vague term used when the plant does not respond after inoculation with the pathogenic agent. Examples: the action of genes *Cf*-2 and *Cf*-9 against *Fulvia fulva*.

In describing these types of reaction, the question has not of course been answered, but this is what you can see with the naked eye or under a microscope; the actual mechanisms are usually unknown. However, genetic resistance is available against the following parasites (**Key:** ★ = resistant varieties available commercially; ■ = selection for resistance in progress; □ = research in progress):

Fungi
- ★ *Fusarium oxysporum* f.sp. *lycopersici*: genes I and I-2
- ★ *Fusarium oxysporum* f.sp. *radicis-lycopersici*: a dominant gene
- ★ *Verticillium dahliae* V. *albo-atrum*: gene *Ve*
- ★ *Pyrenochaeta lycopersici*: gene *pyl*.
- ★ *Phytophthora infestans*: gene *Ph*-2
- ★ *Cladosporium fulvum*: (= *Fulvia fulva*): genes Cf-2, Cf-4, Cf-5, Cf-6, Cf-9, and others
- ★ *Stemphylium* spp: gene *Sm*

Bacteria
- ★ *Pseudomonas tomato*: gene *Pto*
- ★ *Clavibacter michiganensis* (= *Corynebacterium michiganense*)

Viruses
- ★ Tobacco mosaic virus (TMV): genes *Tm*-1, *Tm*-2, *Tm*-2^2
- □ Cucumber mosaic virus (CMV)

- Potato Y virus (PYV)
- Tomato yellow leaf curl (TYLCV)

Nematodes
- ⋆ *Meloidogyne incognita* and others: gene *Mi*

Insects
- *Trialeurodes vaporariorum*
- *Liriomyza* spp

Known resistances of varieties cultivated in the Mediterranean basin

These have been separated into two lists:
- Varieties which breed true from seed
- F1 hybrids

These in turn are divided into two parts:
- Determinate growth
- Indeterminate growth

Key to indication of resistance to disease:

V = *Verticillium dahliae and V. alboatrum*

F = *Fusarium oxysporum* f.sp. *lycopersici* strain 0 (formerly 1)

F2 = *Fusarium oxysporum* f.sp. *lycopersici* strain 0 (formerly 1) and 1 (formerly 2)

Fr = *Fusarium oxysporum* f.sp. *radicis-lycopersici*

N = *Meloidogyne* spp. (nematodes)

P = *Pyrenochaeta lycopersici*

M = *Phytophthora infestans*

C = *Cladosporium fulvum*

S = *Stemphylium* spp.

Pt = *Pseudomonas tomato*

T = Tobacco mosaic virus (=TMV)

Root stocks used for tomatoes and aubergines are listed separately.

True breeding varieties (determinate growth)

Bela	VFS
CalJ	VF
Campbell 1327	VF
Cannery Row	VFS
Chico III	FS
Coudoulet	VF2
Earlymech	VFS
Europeel	VFS
Fline	VF2MS
Flora Dade	VF2S
Heinz 1370	FS
Heinz 1706	VF
Lima	VFS

Macero 2	VF
Marti	VFS
Mecline	VFMS
Mega	VFS
Merkurit	VFS
Peto 94 (= Carlin)	VF2S
Petogro	VF2S
Petomech	VFS
Piline	VF2MS
Rimone	VF2Pt
Rio Fuego	VF2
Rio Grande	VF2
Roma VF	VF
Rossol	VFN
Royal chico VFN	VFN
UC 82	VFS
UC 97-3 (= Pressy)	VF
UC 105	VF
UC 134	VFS
Vesuvio	VF
VF 6203 (= Justar)	VF

True varieties (indeterminate growth)

Earlypack	—
Far	VF
Marmande	—
Marmande VF	VF
Marmande VR	V
Marsol	VFN
Motelle	VF2NS
Piersol	VFN
Raf	F
Saint Pierre	—
San Marzano	—

F1 hybrids (determinate growth)

Aloha	VF2S
Alphamech (= Petopride)	VF2S
Apla	VF2S
Balca	VT
Bandera	VFN
Belote	VF
Caracas	VFNT
Carma	VFN
Count	VF2S
Duck	VF2S
Earlymat	VF2NSPt
Foxy	T
Fusca	VFT
Fusor	VFT
Hypeel 229	VFS
Hypeel 244	VFS
Jackpot	VF2N
Lerica	VF2
Luca	T
Maindor	VF2T
Maritza 25	T
Mecador	VF2S

Nema-mech	VF2NSPt	Fournaise	—
Overpack	FN	Furiak	VF2ST
Precodor	T	Futuria	VFCT
Primosol	VF2	Garanto	VF2CT
Prisca	VFCT	Grinta	T
Quatuor	VT	Hymar	VFN
Safi	F2NS	Kyndia	VFNPCT
Sanzana	VF2S	Larma	F2FrCT
Sunny	VF2S	Lorena	VFNptT
Tetraline	VT	Lucy	T
Topla	VF2S	Madona (= Tanit)	VFNT
Vemar	VFCT	Manific (= Mani)	VFS
Zenith	VF2NSPt	Manon	VF2ST
		Melody	VFT
		Monte Carlo	VFNS

F1 hybrids (indeterminate growth)

Acor	FNT	Montfavet 63–5	—
Alia	VFNT	Nancy	VFST
Amfora	VFCT	Novy	VFNT
Angela	F2CT	Ogosta VFT	VFT
Argus	VF2NCST	Olympe	VFNS
Bali	VMT	Orphee	VFT
Bornia	VFNCT	Perfecto	F2CT
Boulba	VF2NST	Pyrella	VFNPCT
Buffalo	VF2CT	Pyros	NM
Campina	VFNT	Rambo	VF2FrNST
Carmello	VFNST	Ramy	VF2NT
Carpy	VFNCT	Rezano	VFT
Caruso	VF2CT	Rianto	F2CT
Claire	F2CT	Ringo	VFNMT
Cobra	VF2ST	Robin	VFS
Counter	VF2CT	Rody	VF2CT
Cristina	VF2NT	Salima	VF2CT
Dario	VF2NST	Tango	VFMT
Darus	VFNCST	Tarasque	VFN
Diego	VFNT	Tenor	VFNT
Dombito	F2CST	Tirana	V
Dombo	VF2CS	Tresor	VFNT
Dona	VF2NT	Triumph	VFT
Duranto	F2CST	Turquesa	VF2NT
Elcy	VF2NCST	Vemone	CT
Erlidor	VFT	Viga	VF2T
Etna	VFNT	Vivia	VFCT
Faculty 121	FTS		
Fanal	VF		
Fandango	VMT		
Ferline	VF2MS		
Flora	VT		

F1 hybrid rootstocks

KNVF	VFNP
TmKAVF2	VF2NPT
Hires Tm	VFNPT

Appendix 3

Lists of micro-organisms, diseases, pests and parasites, and of illustrations of symptoms and plant damage.

Normal type = diagnosis
Bold type = **descriptions** (biology, treatment)

Micro-organisms

Bacteria

Agrobacterium, sp. 73, 83 (**description 5**)
Bacillus, 129
Clavibacter michiganensis subsp. *michiganensis*, 45, 53, 55, 97, 103, 109, 111, 119, 123 (**description 3**)
Erwinia sp., 97, 105, 119, 129 (**description 5**)
Pseudomonas corrugata, 97, 105, 107, 109, 117 (**description 4**)
Pseudomonas syringae pv. *tomato*, 45, 49, 50, 53, 119, 123 (**description 1**)
Xanthomonas campestris pv. *vesicatoria*, 45, 49, 50, 119, 123 (**description 2**)

Fungi

Acremonium sclerotigenum, 57, 59
Alternaria dauci f.sp. *solani*, 45, 53, 61, 97, 103, 119, 127, 131 (**description 6**)
Alternaria tenuis, Alternaria tenuissima, 119, 125, 131 (**description 12**)
Botryosporium, sp. 97, 103
Botrytis cinerea, 45, 85, 93, 97, 101, 119, 129, 131, 135 (**description 7**)
Cladosporium fulvum, see *Fulvia fulva*
Colletotrichum coccodes, 73, 81, 119, 125 (**descriptions 12 to 14**)
Didymella lycopersici 85, 89, 97, 101 (**description 19**)
Erysiphe sp., 45, 57, 59 (**description 9**)
Fulvia fulva, 45, 57, 59 (**description 8**)
Fusarium oxysporum f.sp. *lycopersici*, 79, 97, 109, 113 (**description 23**)
Fusarium oxysporum f.sp. *radicis-lycopersici*, 73, 79, 85, 89, 97, 109, 118 (**description 16**)
Fusarium sp., 119, 129 (**description 12**)
Geotrichum candidum, 119, 129 (**description 12**)
Hansfordia pulvinata, 57, 59
Leveillula taurica, 45, 57, 59 (**description 9**)
Mucor sp., 119, 129 (*description 12*)
Penicillium sp., 45, 53, 55
Pullularia pullulans, 129 (**description 12**)
Pythium sp., 73, 77, 79, 129 (**description 17**)
Phytophthora sp., 73, 77, 79 (**description 17**)
Phytophthora infestans, 45, 65, 67, 71, 97, 105, 119, 127, 137, 139 (**description 10**)
Phytophthora nicotianae var. *parasitica*, 65, 77, 85, 89, 119, 135, 139 (**description 20**)
Pleospora herbarum, 119, 125 (see also *Stemphylium* spp.) (**description 12**)
Pyrenochaeta lycopersici, 73, 81, 83, 95 (**description 13**)
Rhizoctonia solani, 73, 85, 89, 119, 135 (**description 17 to 21**)
Rhizopus nigricans, 119, 129 (**description 12**)
Sclerotinia sclerotiorum, 85, 93, 97, 101 (**description 22**)
Spongospora subterranea 73, 83 (**description 15**)
Stemphylium spp. 45, 53 (see also *Pleospora herbarum*) (**descriptions 11 and 12**)
Verticillium dahliae, 97, 109, 115 (**description 24**)
Verticillium albo-atrum, see *Verticillium dahliae*

Mycoplasma

Mycoplasma of the Aster yellows group, 15, 19, 27, 37 (**description 30**)

Viruses

Rhabdovirus, 15, 19, 25, 119, 141
Tomato yellow leaf curl, 15, 19, 25, 27, 39 (**description 29**)
Cucumber mosaic virus, 15, 19, 21, 25, 29, 33, 37, 71, 105, 119, 135, 137 (**description 26**)
Alfalfa mosaic virus, 29, 33, 35, 71, 105, 119, 135, 137 (**description 28**)
Tobacco mosaic virus, 15, 21, 25, 29, 33, 51, 119, 137, 141 (**description 25**)
Potato virus Y, 29, 33, 35, 45, 49, 54 (**description 27**)

Parasitic diseases

Bacterial diseases
Bacterial canker, 45, 53, 55, 97, 103, 109, 111, 119, 123 (**description 3**)

Black blight, 97, 105, 107, 109, 117 (**description 4**)
Bacterial gall, 45, 49, 50, 119, 123 (**description 2**)
Specks, 45, 49, 50, 53, 119, 123 (**description 1**)

Fungal diseases

Alternariosis (target spot, early blight), 45, 53, 61, 97, 103, 119, 127, 131 (**description 6**)
Fruit alternariasis (fruit rot), 119, 125, 131 (**description 12**)
Grey mould, 45, 85, 93, 97, 101, 119, 129, 131, 135 (**description 7**)
Anthracnosis, 119, 125 (**description 14**)
Didymella cankers, 85, 89, 97, 101 (**description 19**)
Oïdium (powdery mildew), 45, 57, 79 (**description 9**)
Cladosporiasis (leaf mould), 45, 57, 59 (**description 8**)
Fusarium wilt, 79, 97, 109, 113 (**description 23**)
Root and collar fusariasis (crown and root rot), 73, 79, 85, 89, 97, 109, 118 (**description 16**)
Fruit rot, 119, 125, 129, 131 (**description 12**)
Various root changes, 73, 77, 79 (**description 17**)
'Ground' mildew, 65, 77, 85, 89, 119, 135, 137, 139 (**description 20**)
'Air' mildew, 45, 65, 67, 71, 97, 105, 119, 127, 137, 139 (**description 10**)
Corky root disease, 73, 81, 83, 95 (**description 13**)
Sclerotinia canker, 85, 93, 97, 101 (**description 22**)
Stemphyliosis (Stemphylium leaf spot), 45, 53 (**description 11**)
Verticillium wilt, 97, 109, 115 (**description 24**)

Mycoplasmosis

Stolbur, 15, 19, 27, 29, 37 (**description 30**)

Viral diseases

Form leaf, mottling and necrosis of the tomato, 15, 19, 21, 25, 29, 33, 35, 71, 105, 119, 135, 137 (**description 26**)
Internal browning, 119
Tobacco mosaic virus, 15, 21, 25, 29, 33, 51, 119, 137, 141 (**description 25**)
Necrotic tobacco mosaic, 29, 33, 35, 71, 105, 119, 135, 137 (**description 28**)
Tomato yellow leaf curl, 15, 19, 25, 27, 39 (**description 29**)
Tomato necrotic spots and mosaic, 29, 33, 35, 45, 49, 54 (**description 27**)
Rhabdovirus diseases, 15, 19, 25, 119, 141

Non-parasitic or physiological diseases

Climatic or cultural accidents, 119, 148
Non-parasitic diseases (brown spots on leaves), 45, 61, 64
Collar suffocation, 85, 95
Root suffocation, 73, 77, 79, 133
Silvering, 29, 36 (**description 31**)
Food deficiencies, 29, 37, 38, 39, 40, 41
Corky peduncular scars, 119, 131
Corky stylar scar or catface, 119, 133, 143
'Yellow' collar, 119, 141 (**description 31**)
Sunburn, 119, 127
Vibrator damage, 119, 146
Frost damage, 29, 37, 39, 133, 143, 147
Hail damage, 97, 103, 119, 123, 140
Leaf roll, 15, 25
Excess salinity, 73, 77, 85, 95, 133
Growth cracks, 119, 145
Fruit Pox, 119, 145
Internal browning, 119, 141
Oedema, 15, 25, 45, 53 (**description 31**)
Blotchy ripening, 119, 141 (**description 31**)
Sterile mutants, 15, 21, 23 (**description 31**)
Blossom end rot, 38, 119, 133
Various phytotoxicities, 15, 23, 29, 39, 41, 45, 49, 53, 54, 67, 71, 96, 105, 119, 133, 143, 145 (**description 31**)
Non-parasitic problem (internal browning of fruit), 119, 147
Non-parasitic problem (brown spots on leaves), 64
Russeting (corky epidermis), 119, 146

Pests and parasitic plants

Aculops lycopersici (tomato russet mite), 67, 71, 105 (**Appendix 1**)
Spider mite damage (**Appendix 1**)
Whitefly damage (**Appendix 1**)
Leaf miner damage (**Appendix 1**)
Caterpillar damage (**Appendix 1**)
Bird damage, 119, 146
Aphid damage, 15, 25 (**Appendix 1**)
Bug damage, 119, 147
Meloidogyne spp. 73, 83 (**description 18**)
Cuscuta spp. (**Appendix 1**)
Orobanche spp. (**Appendix 1**)

Illustrations of symptoms caused by micro-organisms

(Micro-organisms responsible or otherwise for parasitic diseases)

Bacteria

Clavibacter michiganensis subsp. *michiganensis*, **93, 124, 126, 193, 209, 212, 214, 216, 217, 243, 296**
Erwinia sp., **185, 197, 198**

Illustrations showing parasitic disease symptoms

Bacterial diseases

Illustrations showing non-parasitic or physiological disease symptoms

Illustrations showing pest and parasitic plant damage

Main works of reference

Diseases of greenhouse plants. Fletcher, J.T. Longman Inc., New York, 1984

Identifying diseases of vegetables. MacNab, A.A., Sherf, A.F., Springer, J.K. Pennsylvania State University, 1983.

Integrated pest management for tomatoes. University of California, Agricultural Sciences Publication 3274, 1982.

Les maladies des plantes maraicheres. Messiaen, C.M., Lafon, R. I.N.R.A., Publ. 6–70, 1970.

Maladies et accidents culturaux de la tomate. Blancard, D., et al. Publication C.T.I.F.L., 1984.

Market diseases of tomatoes, peppers and eggplants. McCulloch, L.P., Cook, H.T., Wright, W.R. U.S.D.A. Agricultural Handbook No. 28, 1968.

Tomato diseases. McKeen, C.D. Canada Department of Agriculture Publication 1479, 1973.

Tomato diseases and their control. Barksdale, T.H., Good, J.M., Danielson, L.L. U.S.D.A. Agricultural Handbook No. 203, 1972.

Tomato diseases. An illustrated guide to their recognition and control. McKay, R. At the Sign of the Three Candles, Fleet Street, Dublin, 1949.

Tomato diseases. A practical guide for seedsmen, growers and agricultural advisors. Watterson, J.C. Petoseed Co., Inc., 1985.

Index

All numbers are page numbers.